a LANGE medical book

Physiology
A Study Guide
Third Edition

William F. Ganong, MD

Jack and DeLoris Lange Professor of Physiology
Department of Physiology
University of California
San Francisco

APPLETON & LANGE
Norwalk, Connecticut/San Mateo, California

0-8385-7875-6

89 90 91 92 93 / 10 9 8 7 6 5 4 3 2 1

Prentice-Hall International (UK) Limited, *London*
Prentice-Hall of Australia, Pty. Limited, *Sydney*
Prentice-Hall Canada, Inc., *Toronto*
Prentice-Hall Hispanoamericana, S.A., *Mexico*
Prentice-Hall of India Private Limited, *New Delhi*
Prentice-Hall of Japan, Inc., *Tokyo*
Simon & Schuster Asia Pte. Ltd., *Singapore*
Editora Prentice-Hall do Brasil Ltda., *Rio de Janeiro*
Prentice-Hall, *Englewood Cliffs, New Jersey*

ISBN: 0-8385-7875-6

ISSN: 0894-2420

Production Editor: Laura K. Giesman
Designer: Steven M. Byrum

PRINTED IN THE UNITED STATES OF AMERICA

Table of Contents

Preface

This *Study Guide* complements the 14th (1989) edition of *Review of Medical Physiology* and is designed to make that book more useful. It can, however, be used with other editions and other books. Students taking a college or professional course in physiology for the first time can use it to organize their thinking, test their knowledge, and prepare for examinations. Physicians, other health professionals, and graduate students preparing for various examinations can use it for review and practice on test questions.

Each of the 39 chapters of this *Study Guide* provide a one-paragraph summary, a list of objectives, a number of general questions, a collection of multiple-choice questions, and a few selected references.

The **objectives** are purposely stated in broad and general terms. They should not be considered a catalog of all the important information in physiology, but they should help students think about much of this information.

The **general questions** should serve a similar purpose and should help students prepare to answer essay questions.

The **multiple-choice questions** are modeled after questions asked on various certifying and Board examinations, including those of the National Board of Medical Examiners and FLEX. The multiple-choice multiple-answer questions are in the same "K question" format used by the National Board of Medical Examiners.

The questions are intentionally graded in terms of difficulty from the simplest to the most complex. A serious effort has been made to include problem-solving questions, to eliminate ambiguity, and to avoid trivia. The questions do not necessarily draw only on the chapter in which they are presented—some are synthesizing questions based on information from various chapters, and a few also require information from other books and references. A key listing of what I consider to be correct **answers to the multiple-choice questions** is provided at the end of the Study Guide.

The **references** are to useful reviews and basic texts that will help students understand the material in the chapters. In some instances, a reference is provided to clinical applications of the material when such applications can help the student understand the physiology. In general, the references do not duplicate but instead complement those at the end of each section of *Review of Medical Physiology,* 14th edition; those in the *Study Guide* are comprehensive, whereas those in *Review of Medical Physiology* are selected to provide background information and details on subjects that are only summarized in the text.

In this third edition, new questions have been added and some older ones deleted. Many of the new questions highlight recent advances that are covered in the 14th edition of *Review of Medical Physiology*.

In addition, more questions that involve illustrations and plotting graphs have been added. The graph questions are designed to help those who have had limited experience with this type of quantitative analysis and presentation. Finally, the wording of some questions has been improved and a few errors have been corrected.

It should be emphasized that as physiologic knowledge advances, it is to a large degree the answers rather than the questions that change, and the answers listed as correct in this edition of the *Study Guide* may be incorrect in the future. In this vein, I welcome comments, criticisms, and suggestions about the questions and about any other part of the *Study Guide*. Please address correspondence to

Dr. William F. Ganong
Department of Physiology, S–762
University of California
San Francisco
CA 94143-0444, USA.

In preparing this third edition of the *Study Guide,* I am particularly indebted to various readers, who helped improve the questions in several chapters, and to Janet Fasbinder and my wife, Ruth, who typed the corrections and additions. Their contributions are greatly appreciated.

William F. Ganong, MD

San Francisco
July, 1989

The General & Cellular Basis of Medical Physiology

1

INTRODUCTION

This study guide is organized in terms of the 39 chapters of the 14th edition of *Review of Medical Physiology*. However, since each chapter of the *Study Guide* starts with a summary of the topics covered, it can also be used by those who do not have a copy of *Review of Medical Physiology* at hand.

For the convenience of the reader, a key listing answers to the multiple-choice questions is provided at the back of the book.

Chapter 1 of *Review of Medical Physiology* is in large part a review of the cellular, molecular, and general basis of medical physiology. The structure of cells and the ways in which cells carry out their various functions are discussed, along with the ways cells communicate with one another. The size and composition of the various body fluid compartments are reviewed, along with the ways in which substances move from one compartment to another and the units used to measure these substances. Finally, the concepts of pH, buffering, homeostasis, and aging are discussed.

OBJECTIVES

The material in the chapter should help students to—

- Name the different fluid compartments in the body, the size of each, and ways their sizes can be measured.

- Define moles, equivalents, and osmoles.

- Compare the various passive and active forces producing movement of substances across cell membranes.

- Define osmosis, and give examples of its role in moving fluid from one location to another in the body.

- Describe and give examples of the coupling of active sodium transport to the movement of other electrolytes and nonelectrolytes across cell membranes in the body.

- Explain resting membrane potential.

- List the unique features of the capillary wall.

- Know the various parts of cells and the functions of each.

- Know the chemical nature and the physiologic significance of the compounds that make up the cell membrane.

- Understand in general terms the structure of DNA and RNA and the role these nucleotides and other substances in the cell play in the process of protein synthesis.

- Define the processes of exocytosis and endocytosis, and describe the contribution each makes to normal cell function.

- Describe the principal ways that the chemical messengers in the extracellular fluid produce changes inside cells.

- List the principal types of receptors.

- Define homeostasis, and give examples of homeostatic mechanisms.

GENERAL QUESTIONS

1. Name the processes responsible for the movement of substances across cell membranes. Which of these require the input of energy? How is this energy provided?

2. Why is Na^+-K^+ ATPase important in physiology?

3. Genes dictate the formation of specific proteins. What are the steps between DNA and the formation of a protein that is secreted by the cell?

4. Why do red blood cells swell and eventually burst when they are placed in a solution of 0.3% sodium chloride?

5. The body fluid compartments are shown in Fig 1–1.
   a. Blood plasma comprises what percentage of body weight?
   b. Interstitial fluid comprises what percentage of body weight?
   c. Intracellular fluid comprises what percentage of body weight?
   d. What separates interstitial fluid from intracellular fluid?
   e. What separates blood plasma from interstitial fluid?

6. Discuss the functions of the proteins found in cell membranes.

7. What is the function of calmodulin? Why is this function important?

8. What happens to resting membrane potential when the extracellular K^+ concentration is increased from 5 meq/L to 10 meq/L? What happens when the extracellular Na^+ concentration is increased from 142 meq/L to 155 meq/L? Why?

9. What is receptor-mediated endocytosis? Discuss its role in the function of the cell.

10. Compare the composition of plasma, interstitial fluid, and intracellular fluid. Explain the differences.

11. Discuss aging from the point of view of its cellular and molecular bases, its physiologic consequences in humans, and its implications in the practice of medicine.

MULTIPLE-CHOICE QUESTIONS

In the following questions, select the single best answer.

1. Cell membranes—
   a. consist almost entirely of protein molecules.
   b. are impermeable to fat-soluble substances.

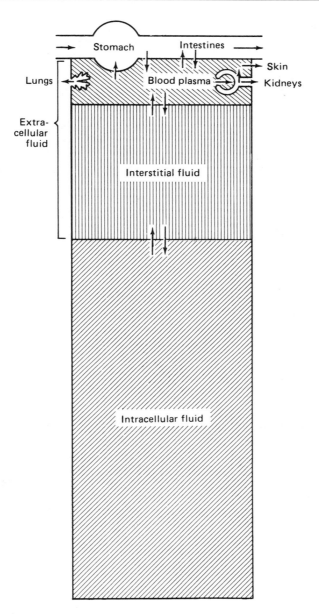

Figure 1–1. Body fluid compartments. Arrows represent fluid movement. Transcellular fluids, which constitute a very small percentage of total body fluids, are not shown. (Modified and reproduced, with permission, from Gamble JL: *Chemical Anatomy, Physiology, and Pathology of Extracellular Fluid,* 6th ed. Harvard Univ Press, 1954.)

    c. in some tissues permit the transport of glucose at a greater rate in the presence of insulin.
    d. are freely permeable to electrolytes but not to proteins.
    e. are stable throughout the life of the cell.

2. Water molecules move from the blood plasma to the interstitial fluid (see Fig 1–1) by—
    a. active transport.
    b. cotransport with H^+.
    c. facilitated diffusion.
    d. cotransport with Na^+.
    e. endocytosis and exocytosis.

3. Second messengers—
   a. are substances that interact with first messengers inside cells.
   b. are substances that bind to first messengers in the cell membrane.
   c. are hormones secreted by cells in response to stimulation by another hormone.
   d. mediate the intracellular responses to many different hormones and neurotransmitters.
   e. are not formed in the brain.

4. The resting membrane potential of a cell—
   a. is dependent on the permeability of the cell membrane to K^+ being greater than the permeability to Na^+.
   b. falls to zero immediately when Na^+-K^+ ATPase in the membrane is inhibited.
   c. is usually equal to the equilibrium potential for K^+.
   d. is usually equal to the equilibrium potential for Na^+.
   e. is markedly altered if the extracellular Na^+ concentration is increased.

5. Proteins that are secreted by cells are generally—
   a. not synthesized on membrane-bound ribosomes.
   b. initially synthesized with a signal peptide or leader sequence at their C terminal.
   c. "packaged" in the Golgi apparatus.
   d. moved across the cell membrane by endocytosis.
   e. secreted in a form that is larger than the form present in the endoplasmic reticulum.

6. Osmosis is—
   a. movement of solvent across a semipermeable membrane from an area where the hydrostatic pressure is high to an area where the hydrostatic pressure is low.
   b. movement of solute across a semipermeable membrane from an area in which it is in low concentration to an area in which it is in high concentration.
   c. movement of solute across a semipermeable membrane from an area in which it is in high concentration to an area in which it is in low concentration.
   d. movement of solvent across a semipermeable membrane from an area in which it is in low concentration to an area in which it is in high concentration.
   e. movement of solvent across a semipermeable membrane from an area in which it is in high concentration to an area in which it is in low concentration.

In the following questions, one or more than one of the answers may be correct. Select—

> **A** if (1), (2), and (3) are correct;
> **B** if (1) and (3) are correct;
> **C** if (2) and (4) are correct;
> **D** if only (4) is correct; and
> **E** if all are correct.

7. Deuterium oxide and inulin are injected into a normal 30-year-old man. The volume of distribution of deuterium oxide is found to be 42 L and that of inulin 14 L.
   (1) The man's intracellular fluid volume is about 14 L.
   (2) The man's intracellular fluid volume is about 28 L.

(3) The man's plasma volume is about 7 L.
(4) The man's interstitial fluid volume is about 10.5 L.

A   B   C   D   E

8. Which of the following are found in mitochondria?
    (1) lysosomal membranes.
    (2) ATP.
    (3) ribonuclease.
    (4) DNA.

A   B   C   D   E

9. Which of the following receptors span the cell membrane 7 times?
    (1) β-adrenergic receptor.
    (2) rhodopsin.
    (3) 5-HT$_{IC}$ receptor.
    (4) mineralocorticoid receptor.

A   B   C   D   E

10. Which of the following receptors have homologous DNA-binding do-
    mains?
    (1) progesterone receptor.
    (2) triiodothyronine receptor (α or β).
    (3) glucocorticoid receptor.
    (4) M$_1$ muscarinic cholinergic receptor.

A   B   C   D   E

11. The rate of diffusion of a solute across a membrane depends on the—
    (1) total area of membrane exposed to the solute.
    (2) thickness of the membrane.
    (3) temperature of the solvent.
    (4) concentration gradient for the solute across the membrane.

A   B   C   D   E

12. When fluid shifts from the plasma to the interstitial space, there is an
    increase in the—
    (1) plasma protein concentration.
    (2) plasma glucose concentration.
    (3) hematocrit.
    (4) plasma Na^+ concentration.

A   B   C   D   E

13. Which of the following act intracellularly to produce physiologic effects?
    (1) triiodothyronine.
    (2) inositol triphosphate.
    (3) aldosterone.
    (4) cyclic AMP.

A   B   C   D   E

14. In which of the following responses do G proteins play a role?
(1) increased cytoplasmic free Ca^{2+}.
(2) increased cyclic AMP.
(3) decreased cyclic AMP.
(4) opening of K^+ channels.

A   B   C   D   E

In the following questions, match each numbered item with the lettered items below that are closely associated with it. Lettered items may be selected once, more than once, or not at all.

(A) deficiency of nucleotide regulatory (G) protein.
(B) antibodies against receptors.
(C) abnormal receptor for extracellular protein.
(D) deficiency of receptors for extracellular protein.

15. Many cases of Graves' disease   A B C D

16. Some cases of pseudohypoparathyroidism   A B C D

17. Some cases of diabetes mellitus   A B C D

18. Some cases of familial hypercholesterolemia   A B C D

In the following questions, indicate whether the item on the left is greater than (G), the same as (S), or less than (L) the item on the right.

19. Hydrogen ion concentration in lysosomes.   G S L   Hydrogen ion concentration in cytoplasm of cells.

20. Plasma osmolality 30 minutes after intravenous infusion of 1000 mL of isotonic (0.9%) sodium chloride.   G S L   Plasma osmolality 30 minutes after intravenous infusion of 1000 mL of isotonic (5%) glucose.

21. Extracellular fluid volume 30 minutes after intravenous infusion of isotonic (5%) glucose.   G S L   Extracellular fluid volume 30 minutes after intravenous infusion of isotonic (0.9%) sodium chloride.

22. Total body water 30 minutes after intravenous infusion of 1000 mL of isotonic (5%) glucose.   G S L   Total body water 30 minutes after intravenous infusion of 1000 mL of isotonic (0.9%) sodium chloride.

23. Contribution of normal concentrations of plasma glucose to total plasma osmolality.   G S L   Contribution of normal concentrations of plasma Na^+ to total plasma osmolality.

24. Calculated plasma volume when some of the dye used to measure it is unknowingly injected subcutaneously instead of intravenously.   G S L   Calculated plasma volume when all of the dye used to measure it is injected intravenously.

25. Concentration of K^+ in intracellular fluid.   **G S L**   Concentration of K^+ in interstitial fluid.

26. Concentration of K^+ in plasma.   **G S L**   Concentration of K^+ in interstitial fluid.

27. Concentration of Ca^{2+} in intracellular fluid.   **G S L**   Concentration of Ca^{2+} in interstitial fluid.

REFERENCES

Alberts B et al: *Molecular Biology of the Cell.* Garland, 1983.

Andreoli T, Hoffman JF, Fanestil DD (editors): *Physiology of Membrane Disorders.* Plenum, 1978.

2 Excitable Tissue: Nerve

Chapter 2 of *Review of Medical Physiology* is the first of 4 concerned with nerves and muscles and the interactions between them. It deals with the properties of neurons: their morphology and the electrical and ionic events that underlie their excitation and their ability to conduct impulses. The various nerve fiber types found in peripheral nerves are also discussed, along with nerve growth factor and glial cells.

OBJECTIVES

The material in the chapter should help students to—

- Name the various parts of a neuron and the functions of each.

- Distinguish between unmyelinated and myelinated neurons, describe the chemical nature of myelin, and summarize the differences in the way unmyelinated and myelinated neurons conduct.

- Define orthograde axoplasmic transport, retrograde axoplasmic transport, and wallerian degeneration.

- Define excitation, electrotonic potential, action potential, and conduction, and describe the changes in membrane permeability and ionic movements that underlie each of these phenomena.

- Explain the following characteristics of a nerve impulse: latent period, firing level, spike potential, after-depolarization, and after-hyperpolarization.

- Describe and explain the compound action potential of mixed nerves.

- List the various nerve fiber types, and comment on their significance in terms of normal and abnormal function of peripheral nerves.

- List the subunits of the nerve growth factor molecule, and comment on the function of each.

- Describe the various types of glial cells.

GENERAL QUESTIONS

1. Of what evolutionary benefit was the development of myelinated neurons in complex multicellular animals and vertebrates?

2. Compare the function of dendrites and the function of axons.

3. What are the electrical and underlying ionic events that lead to a propagated action potential? What makes the action potential move away from the site of excitation?

4. Describe and explain the effects of tetrodotoxin on the resting membrane potential and the action potential of nerve cells.

5. What nerve fiber types would you expect to find in a sympathetic nerve from the celiac ganglion to the intestine? Explain your answer.

6. What are the origins of the various types of glia? Discuss the function of each type.

7. What is the "all or none" law? What is its physiologic basis?

8. What are the equilibrium potentials for Na^+, K^+, and Cl^- in nerves? What is the significance of the difference between the equilibrium potentials of the various ions and the resting membrane potential?

9. How do the absolute and relative refractory periods correlate in time with the various phases of the action potential? Why is the nerve refractory during these periods?

10. Compare the electrical and ionic events in neurons to those in skeletal, cardiac, and smooth muscle.

11. What types of ion channels are found in the cell membranes of neurons? What is the function of each type?

MULTIPLE-CHOICE QUESTIONS

In the following questions, select the single best answer.

1. The action potential of a neuron—
   a. is initiated by efflux of Na^+.
   b. is terminated by efflux of K^+.
   c. declines in amplitude as it moves along the axon.
   d. results in a transient reversal of the concentration gradient of Na^+ across the cell membrane.
   e. is not associated with any net movement of Na^+ or K^+ across the cell membrane.

2. When stimulating electrodes are placed on the external surface of a single unmyelinated axon and current is passed—
   a. if the stimulating current is of threshold intensity, action potentials are produced that move away from the cathode in both directions.
   b. if the stimulating current is subthreshold, the membrane potential is greater at the cathode than at the anode.
   c. if the stimulating current is subthreshold, a local potential is generated that moves away from the cathode in both directions.
   d. if the stimulating current is of threshold intensity, a monophasic action potential will be recorded by recording electrodes on the external surface of the axon 10 cm from the cathode.
   e. if the stimulating current is of threshold intensity, the excitability of the axon to subsequent stimulation at the anode will be increased for several milliseconds.

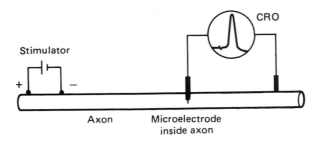

Figure 2–1.

3. A squid axon is placed on stimulating electrodes, and an intracellular electrode is inserted and connected through a cathode-ray oscilloscope (CRO) to an indifferent electrode (Fig 2–1). When the axon is stimulated, the latent period is 1.5 ms. The intracellular electrode is 6 cm from the anode of the stimulator and 4.5 cm from the cathode of the stimulator. What is the conduction velocity of the axon?
   a. 15 m/s.
   b. 30 m/s.
   c. 40 m/s.
   d. 67.5 m/s.
   e. this cannot be determined from the information given.

4. Which of the following has the slowest conduction velocity?
   a. A α fibers.
   b. A β fibers.
   c. A γ fibers.
   d. B fibers.
   e. C fibers.

5. A man falls into a deep sleep with one arm under his head. This arm is paralyzed when he awakens, but it tingles, and pain sensation in it is still intact. The reason for the loss of motor function without loss of pain sensation is that—
   a. A fibers are more susceptible to hypoxia than B fibers.
   b. A fibers are more sensitive to pressure than C fibers.
   c. C fibers are more sensitive to pressure than A fibers.
   d. sensory nerves are nearer the bone than motor nerves and hence are less affected by cooling.
   e. sensory nerves arc nearer the bone than motor nerves and hence are less affected by pressure.

6. Which part of a neuron has the highest concentration of Na^+ channels per square millimeter of cell membrane?
   a. dendrites.
   b. cell body near dendrites.
   c. synaptic knobs.
   d. axonal membrane under myelin.
   e. none of the above.

In the following questions, one or more than one of the answers may be correct. Select—

   A if (1), (2), and (3) are correct;
   B if (1) and (3) are correct;
   C if (2) and (4) are correct;
   D if only (4) is correct; and
   E if all are correct.

7. In a typical spinal motor neuron—
   (1) protein synthesis occurs in the cell body rather than in the endings.
   (2) substances are transported inside the axon from the cell body to the nerve endings.
   (3) the axon is much longer than the dendrites.
   (4) substances are transported inside the axon from the nerve endings to the cell body.

   A   B   C   D   E

8. Saltatory conduction—
   (1) occurs only in myelinated neurons.
   (2) is slower than nonsaltatory conduction.
   (3) stops if a local anesthetic is applied to a node of Ranvier.
   (4) does not occur with antidromic conduction.

   A   B   C   D   E

9. Nerve growth factor—
   (1) is made up of 3 polypeptide subunits.
   (2) is found in high concentration in the submaxillary salivary glands of male mice.
   (3) is necessary for the growth and development of the sympathetic nervous system.
   (4) is picked up by nerves from the organs they innervate.

   A   B   C   D   E

10. In a typical spinal motor neuron—
    (1) the action potential is initiated in the axon hillock region.
    (2) an antidromically conducted stimulus invades the soma and proximal dendrites but does not depolarize presynaptic terminals ending on these structures.
    (3) the axon usually divides into 2 or more branches.
    (4) action potentials are reduced in size at each branch.

    A   B   C   D   E

11. Which of the following are actively transported out of neurons?
    (1) Ca^{2+}.
    (2) K^+.
    (3) Na^+.
    (4) HPO_4^{2-}.

    A   B   C   D   E

In the following questions, indicate whether the item on the left is greater than (G), the same as (S), or less than (L) the item on the right.

12. Increase in Na^+ permeability during the rising phase of the action potential.   G S L   Increase in K^+ permeability during the rising phase of the action potential.

13. Rate of conduction in small-diameter axons.   G S L   Rate of conduction in large-diameter axons.

14. Rate of conduction in somatic nerve that innervates
skeletal muscle.
G S L
Rate of conduction in sympathetic nerve that innervates smooth muscle.

15. Excitability of neuron when
membrane potential is in-
creased to −80 mV.
G S L
Excitability of neuron when
membrane potential is re-
duced to −60 mV.

16. Excitability of neuron during overshoot portion of
action potential.
G S L
Excitability of neuron during after-depolarization
portion of action potential.

17. Total duration of compound action potential of a
mixed nerve 5 mm from
stimulating electrode.
G S L
Total duration of com-
pound action potential of a
mixed nerve 50 mm from
stimulating electrode.

18. Myelination of axons that
mediate "fast" pain.
G S L
Myelination of axons that
mediate "slow" pain.

19. Excitability of neuron during electrotonic potential at
anode.
G S L
Excitability of neuron during electrotonic potential at
cathode.

20. Size of compound action
potential due to B fibers in
dorsal root of spinal nerve.
G S L
Size of compound action
potential due to B fibers in
ventral root of spinal
nerve.

REFERENCES

Kuffler SW, Nicholls JG, Martin R: *From Neuron to Brain,* 2nd ed. Sinauer Associates, 1984.
Stevens CF: The neuron. *Sci Am* (Sept) 1979;**241**:54.

Excitable Tissue: Muscle

3

Chapter 3 of *Review of Medical Physiology* outlines the morphologic and functional characteristics of the 3 types of muscle found in the body: skeletal, cardiac, and smooth muscle. Actin, myosin, and the other contractile proteins are discussed in terms of the contraction of all 3 types of muscle and in terms of the structural and contractile roles of these proteins in many other cells.

OBJECTIVES

The material in the chapter should help students to—

- Describe the gross and microscopic anatomy of skeletal muscle, including the cross-striations, the relation of actin to myosin, and the sarcotubular system.

- List the sequence of events leading from an action potential in the motor nerve to contraction of a skeletal muscle, and discuss the significance of each.

- Compare isometric and isotonic contractions.

- Explain summation of contractions and the relation between initial muscle length and strength of contraction.

- Describe the sources of energy for muscle contraction, and explain how energy is transferred to the contractile mechanism.

- Define oxygen debt, and describe its role in muscle function during exercise.

- Describe the differences between fast and slow skeletal muscles.

- Define the term motor unit, and discuss the effects of denervation on skeletal muscle.

- Analyze the morphologic differences between cardiac and skeletal muscle.

- Describe the extracellularly and intracellularly recorded action potential of cardiac muscle and its relation to the contractile response of the muscle and the electrocardiogram.

- Outline the ionic events that underlie the various phases of the action potential of cardiac muscle.

- Describe the fluctuations in membrane potential seen in pacemaker tissue and the ionic events responsible for the prepotential.

- Compare visceral smooth muscle and multi-unit smooth muscle to skeletal muscle.

- Summarize the effects of acetylcholine and norepinephrine on the electrical and contractile activity of visceral smooth muscle.

GENERAL QUESTIONS

1. Discuss excitation-contraction coupling in skeletal muscle, and compare it to excitation-contraction coupling in cardiac and smooth muscle.

2. Define total tension, passive tension, active tension, resting length, and equilibrium length.

3. Cells in the conducting tissue of the heart and certain neurons in the brain (eg, those responsible for respiration) discharge spontaneously in a rhythmic fashion. What is the feature common to all these cells, how is it produced, and how can the rate of discharge of these cells be altered?

4. "Muscle is a machine for converting chemical into mechanical energy." Analyze and discuss this statement.

5. What is the minimum stimulation frequency at which tetanus occurs in a fast muscle with a twitch duration of 7.5 ms, and in a slow muscle with a twitch duration of 100 ms?

6. Discuss the role of Ca^{2+} in muscle contraction. How does it relate to treppe?

7. Discuss the factors that permit gradation of skeletal muscle responses in a living, intact animal.

8. In living animals, what determines the initial length (resting length) of the muscle fibers in the heart, and how do alterations in developed tension manifest themselves? Discuss the role of the length-tension relationship in cardiac muscle in the normal and abnormal function of the heart.

9. Compare the mechanism that initiates contraction in smooth muscle with the mechanism that initiates contraction in skeletal muscle. What are the similarities and the differences?

10. Discuss tone. Does it occur in all 3 types of muscle?

11. How does the relationship of actin to myosin explain the length-tension curve of skeletal muscle?

12. Define plasticity as the term is applied to muscle. What are the practical physiologic consequences of this property of smooth muscle?

MULTIPLE-CHOICE QUESTIONS

In the following questions, select the single best answer.

1. The action potential of skeletal muscle—
   a. has a prolonged plateau phase.
   b. spreads inward to all parts of the muscle via the T tubules.
   c. causes the immediate uptake of Ca^{2+} into the lateral sacs of the sarcoplasmic reticulum.
   d. is longer than the action potential of cardiac muscle.
   e. is not essential for contraction.

2. The functions of tropomyosin in skeletal muscle include—
   a. sliding on actin to produce shortening.
   b. releasing Ca^{2+} after initiation of contraction.
   c. binding to myosin during contraction.

d. acting as a "relaxing protein" at rest by covering up the sites where myosin binds to actin.

e. generating ATP, which it passes to the contractile mechanism.

3. The contractile response in skeletal muscle—
   a. starts after the action potential is over.
   b. does not last as long as the action potential.
   c. produces more tension when the muscle contracts isotonically than when the muscle contracts isometrically.
   d. produces more work when the muscle contracts isometrically than when the muscle contracts isotonically.
   e. can increase in magnitude with repeated stimulation.

4. Gap junctions—
   a. are absent in cardiac muscle.
   b. are present but of little functional importance in cardiac muscle.
   c. are present and provide the pathway for rapid spread of excitation from one cardiac muscle fiber to another.
   d. are absent in smooth muscle.
   e. connect the sarcotubular system to individual skeletal muscle cells.

5. A motor unit is made up of—
   a. a flexor muscle and an extensor muscle.
   b. a single skeletal muscle and all the motor neurons that supply it.
   c. a single motor neuron and all the muscle fibers it innervates.
   d. a large bundle of muscle fibers.
   e. all the motor neurons in which responses are observed after maximal stimulation of a single sensory nerve.

6. The cross-bridges of the sarcomere in skeletal muscle are components of—
   a. actin.
   b. myosin.
   c. troponin.
   d. tropomyosin.
   e. myelin.

7. Intestinal smooth muscle—
   a. does not contain tight junctions.
   b. contracts when Ca^{2+} is taken up by the sarcoplasmic reticulum.
   c. generally responds by contracting when norepinephrine is added to it in vitro.
   d. contracts when actin and myosin filaments shorten.
   e. contracts when stretched in the absence of any extrinsic innervation.

8. The reason a skeletal muscle can maintain a voluntary submaximal contraction for long periods without a decrease in tension due to fatigue is—
   a. treppe.
   b. summation of contractions.
   c. elasticity of components that are in parallel with the contractile components.
   d. asynchronous firing of motor units.
   e. increased consumption of lactic acid.

In the following questions, select—

   A if the item is associated with **(a)** below,
   B if the item is associated with **(b)** below,
   C if the item is associated with both **(a)** and **(b)**, and
   D if the item is associated with neither **(a)** nor **(b)**.

(a) Strength of contraction of a given muscle.
(b) Duration of contraction of the muscle.

9. Length of muscle at start of contraction   **A**   **B**   **C**   **D**

10. Type of myosin heavy chain isoform in muscle   **A**   **B**   **C**   **D**

11. Degree to which firing in motor nerve to muscle is asynchronous   **A**   **B**   **C**   **D**

12. Troponin content of muscle   **A**   **B**   **C**   **D**

13. Dystrophin content of muscle   **A**   **B**   **C**   **D**

In the following questions, one or more than one of the answers may be correct. Select—

    A if (1), (2), and (3) are correct;
    B if (1) and (3) are correct;
    C if (2) and (4) are correct;
    D if only (4) is correct; and
    E if all are correct.

14. The Z lines in muscle—
    (1) move farther apart when the muscle is stretched.
    (2) are clearly visible in skeletal and cardiac muscle.
    (3) move closer together during contraction.
    (4) mark the site where actin and myosin overlap.

    A   **B**   **C**   **D**   **E**

15. Ca^{2+}—
    (1) initiates contraction in skeletal muscle.
    (2) is involved in the mechanism by which digitalis and related drugs increase the strength of contraction of cardiac muscle.
    (3) initiates contraction in cardiac muscle.
    (4) plays a role in the action of norepinephrine on intestinal smooth muscle.

    A   **B**   **C**   **D**   **E**

16. Ca^{2+} concentration is very low (10^{-8} mol/L) in relaxed muscle fibers because—
    (1) Ca^{2+} is consumed during contraction.
    (2) Ca^{2+} is bound to mitochondria.
    (3) Ca^{2+} diffuses down its electrical gradient to the extracellular fluid.
    (4) Ca^{2+} pumps transport free cytoplasmic Ca^{2+} into the sarcoplasmic reticulum and out of the cell.

    A   **B**   **C**   **D**   **E**

17. The following occur in a denervated skeletal muscle—
    (1) fibrillations.
    (2) paralysis.
    (3) appearance of acetylcholine receptors outside the motor end plate.
    (4) increased sensitivity of the muscle to acetylcholine.

    A   **B**   **C**   **D**   **E**

18. Actin is found in—
(1) neurons.
(2) smooth muscle.
(3) fibroblasts.
(4) skeletal muscle.

A   B   C   D   E

19. Which of the following can be used by cardiac muscle as energy sources?
(1) ketones.
(2) amino acids.
(3) free fatty acids.
(4) glycogen.

A   B   C   D   E

In the following questions, indicate whether the item on the left is greater than **(G)**, the same as **(S)**, or less than **(L)** the item on the right.

20. Duration of the increase in Ca^{2+} influx during the action potential of cardiac muscle.   **G   S   L**   Duration of the increase in Na^+ influx during the action potential of cardiac muscle.

21. In a normal man before breakfast, proportion of the caloric needs of the heart obtained from fat.   **G   S   L**   In a normal man before breakfast, proportion of the caloric needs of the heart obtained from carbohydrate.

22. Contractile activity of intestinal smooth muscle after application of norepinephrine to the muscle.   **G   S   L**   Contractile activity of intestinal smooth muscle after application of acetylcholine to the muscle.

23. Tension in wall of bladder 1 second after it is distended.   **G   S   L**   Tension in wall of bladder 1 minute after it is distended to the same degree.

24. Excitability of uterine smooth muscle after administration of progesterone.   **G   S   L**   Excitability of uterine smooth muscle after administration of estrogen.

25. ATPase activity of cardiac muscle after thyroidectomy.   **G   S   L**   ATPase activity of cardiac muscle after thyroxine (T_4) treatment.

REFERENCES

Stracher A (editor): *Muscle and Nonmuscle Motility.* Academic Press, 1983.
Wilkie DR: *Muscle,* 2nd ed. Arnold, 1976.

4 Synaptic & Junctional Transmission

Chapter 4 is concerned with the junctions across which transmission occurs from one neuron to another and from neurons to muscle cells. The anatomic features of the synapses between neurons are summarized, and the stimulatory and inhibitory electrical responses that interact at these junctions—along with their functional significance—are reviewed in detail. The chemical nature, synthesis, and metabolism of synaptic transmitters in all parts of the nervous system are discussed, along with their receptors. The importance of the synapse as a site for physiologic and pharmacologic modification of brain function is emphasized. The unique properties of the neuromuscular junction—a junction which has been studied in great detail—are reviewed. In addition, the general effects of denervation are discussed.

OBJECTIVES

The material in the chapter should help students to—

- Distinguish between chemical and electrical transmission at synapses.

- List the morphologic types of synapses seen in the body.

- Define convergence and divergence in neural networks, and discuss their implications.

- Describe fast and slow excitatory and inhibitory postsynaptic potentials, and outline the permeability changes and ionic fluxes that underlie them.

- Discuss how postsynaptic potentials interact to generate an action potential.

- Define and give examples of direct inhibition, indirect inhibition, presynaptic inhibition, and postsynaptic inhibition.

- Define synaptic plasticity, and discuss its relation to learning.

- Know the types of chemicals that serve as synaptic transmitters.

- List 10 neurotransmitters and the principal sites in the nervous system at which they are released.

- Summarize the steps involved in the biosynthesis, action, and breakdown of acetylcholine, dopamine, norepinephrine, epinephrine, and serotonin.

- Define the term opioid peptide, list the principal opioid peptides in the body, and name the precursor molecules from which they come.

- Describe the myoneural junction, and explain how action potentials in the motor neuron at the junction lead to contraction of the skeletal muscle.

- Define and explain denervation hypersensitivity.

GENERAL QUESTIONS

1. Conduction can occur in either direction along an axon but in only one direction at a chemical synaptic junction. Why?

2. Discuss neuronal regeneration.

3. Release of an excitatory mediator from a terminal button of a presynaptic neuron generally increases the permeability of the adjacent membrane of the postsynaptic neuron to Na^+ and K^+. Why doesn't this increase in permeability regularly produce an action potential in the postsynaptic cell?

4. One of the records in Fig 4–1 was obtained with acetylcholine acting as an excitatory transmitter, and the other was obtained with gamma-aminobutyric acid (GABA) acting as an inhibitory transmitter. Which is which? Explain your answer.

5. What is the role of reuptake in the metabolism of amine neurotransmitters? How does its absence contribute to denervation hypersensitivity? Does it occur in the endings of neurons secreting other types of neurotransmitters?

6. "In the human forebrain, the ratio of synapses to neurons has been calculated to be 40,000 to 1." Discuss this statement and its physiologic implications.

7. What is synaptic delay? Discuss its significance in terms of brain function.

8. Discuss the role of glutamate and the physiology of the glutamate receptors in the brain.

9. Although some neurons secrete neurotransmitters, others secrete hormones, ie, chemical messengers that enter the bloodstream. Name the hormones secreted in this fashion, and identify and locate the neurons involved.

10. The nerve gases that have been developed for chemical warfare generally inhibit acetylcholinesterase. Why does inhibition of acetylcholinesterase produce death?

11. Compare the endings of somatic motor nerves in skeletal muscle to the endings of autonomic motor nerves in cardiac and smooth muscle.

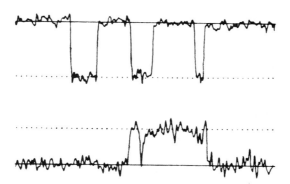

Figure 4–1. Patch clamp records obtained from 2 different neurons. Downward deflection indicates inward current pulses. Upward deflection indicates outward current pulses.

MULTIPLE-CHOICE QUESTIONS

In the following questions, select the single best answer.

1. Which of the following is *not* synthesized in postganglionic sympathetic neurons?
   a. L-dopa.
   b. dopamine.
   c. norepinephrine.
   d. epinephrine.
   e. acetylcholine.

2. Fast inhibitory postsynaptic potentials (IPSPs)—
   a. are produced by the process of indirect inhibition.
   b. are a consequence of presynaptic inhibition.
   c. interact with other fast and slow potentials to move the membrane potential of the postsynaptic neuron toward or away from the firing level.
   d. are produced by a decrease in K^+ permeability.
   e. occur in skeletal muscle.

3. Initiation of an action potential in skeletal muscle by stimulating its motor nerve—
   a. requires spatial facilitation.
   b. requires temporal facilitation.
   c. requires the release of norepinephrine.
   d. is inhibited by a high concentration of Ca^{2+} at the neuromuscular junction.
   e. requires the release of acetylcholine.

4. Fast excitatory postsynaptic potentials (EPSPs)—
   a. are produced by an increase in the permeability of the membrane of the postsynaptic cells to K^+.
   b. usually produce enough depolarization to trigger an action potential in postsynaptic neurons.
   c. are actively propagated potentials.
   d. are produced by acetylcholine acting on the nodal tissues of the heart.
   e. can summate temporally with other fast EPSPs.

5. Which of the following synaptic transmitters is *not* a peptide, polypeptide, or protein?
   a. substance P.
   b. met-enkephalin.
   c. β-endorphin.
   d. serotonin.
   e. dynorphin.

6. Activation of which of the following receptors would be expected to decrease anxiety?
   a. nicotinic cholinergic receptors.
   b. glutamate receptors.
   c. $GABA_A$ receptors.
   d. glucocorticoid receptors.
   e. α_1-adrenergic receptors.

7. Which of the following receptors is coupled to a G protein?
   a. glycine receptor.
   b. $GABA_A$ receptor.

   c. nicotinic cholinergic receptor.
   d. α_2-adrenergic receptor.
   e. glutamate receptor.

In the following questions, one or more than one of the answers may be correct. Select—

        A if (1), (2), and (3) are correct;
        B if (1) and (3) are correct;
        C if (2) and (4) are correct;
        D if only (4) is correct; and
        E if all are correct.

8. GABA is—
   (1) an inhibitory transmitter in the brain.
   (2) synthesized in a reaction catalyzed by glutamate decarboxylase (GAD).
   (3) facilitated in its actions by diazepam (Valium) and other benzodiazepines.
   (4) present in the same neurons as acetylcholine.

        A   **B**   **C**   **D**   **E**

9. Which of the following are known to be associated with other transmitters in neurons, ie, to be cotransmitters?
   (1) norepinephrine.
   (2) dopamine.
   (3) serotonin.
   (4) acetylcholine.

        A   ·**B**   **C**   **D**   **E**

10. After an injection of curare, a drug that blocks nicotinic acetylcholine receptors, one would expect to find—
   (1) a diminished response of the heart to stimulation of the vagus nerves.
   (2) a diminished response of intestinal smooth muscle to acetylcholine.
   (3) a diminished response of skeletal muscles to direct electrical stimulation.
   (4) a diminished response of skeletal muscles to stimulation of their motor nerves.

        A   **B**   **C**   **D**   **E**

11. Which of the following hormones are also neurotransmitters?
   (1) somatostatin.
   (2) oxytocin.
   (3) epinephrine.
   (4) ACTH.

        A   **B**   **C**   **D**   **E**

12. One would expect noradrenergic transmission to be enhanced by drugs that—
   (1) facilitate norepinephrine reuptake.
   (2) inhibit tyrosine hydroxylase.

(3) inhibit dopamine β-hydroxylase.
(4) inhibit monoamine oxidase.

A   **B**   **C**   **D**   **E**

In the following questions, indicate whether the item on the left is greater than **(G)**, the same as **(S)**, or less than **(L)** the item on the right.

13. Amount of norepinephrine released at a noradrenergic synapse when presynaptic α_2-adrenergic receptors are stimulated.   **G S L**   Amount of norepinephrine released at a noradrenergic synapse when postsynaptic α_2-adrenergic receptors are stimulated.

14. Number of acetylcholine receptors on the surface of denervated skeletal muscle before nerve regrows.   **G S L**   Number of acetylcholine receptors on the surface of denervated skeletal muscle after nerve regrows.

15. Amount of neurotransmitter released by a nerve impulse after habituation.   **G S L**   Amount of neurotransmitter released by a nerve impulse after sensitization.

16. Number of times a β_2-adrenergic receptor spans the cell membrane.   **G S L**   Number of times an M_1 muscarinic acetylcholine receptor spans the cell membrane.

17. Amount of acetylcholine in endings of preganglionic sympathetic neurons.   **G S L**   Amount of norepinephrine in preganglionic sympathetic neurons.

18. Quantal release of acetylcholine at rest at a synapse where acetylcholine is an excitatory mediator.   **G S L**   Quantal release of acetylcholine during IPSP in a postsynaptic neuron at the same synapse.

19. Amount of Nissl substance visible in the cell body of a neuron before transection of its axon.   **G S L**   Amount of Nissl substance visible in the cell body of a neuron after transection of its axon.

20. Increase in intracellular cyclic AMP produced by activation of an α_2-adrenergic receptor.   **G S L**   Increase in intracellular cyclic AMP produced by activation of a β_1-adrenergic receptor.

21. Amount of L-dopa in a neuron after administration of a drug that inhibits tyrosine hydroxylase.   **G S L**   Amount of L-dopa in a neuron after administration of a drug that inhibits dopa decarboxylase.

22. Effect of activation of $GABA_A$ receptor on K^+ conductance.   **G S L**   Effect of activation of $GABA_B$ receptor on K^+ conductance.

REFERENCES

Cooper JR, Bloom FE, Roth RH: *The Biochemical Basis of Neuropharmacology,* 5th ed. Oxford Univ Press, 1986.

Shepherd GM: *The Synaptic Organization of the Brain,* 2nd ed. Oxford Univ Press, 1979.

5

Initiation of Impulses in Sense Organs

Chapter 5 lists the various senses, the sensory receptors that initiate action potentials in sensory nerves, and the sense organs that contain the receptors. The way impulses are generated in sensory nerves is then considered. Local generator potentials are described, and the way they depolarize sensory nerve fibers is discussed. Adaptation and the principles and laws that govern the coding of sensory information are also reviewed.

OBJECTIVES

The material in the chapter should help students to—

- Define the term sensory receptor, and distinguish between this type of receptor and the receptors that bind neurotransmitters, hormones, and other ligands.

- Define adequate stimulus, and give examples.

- Enumerate the senses, and discuss the various classifications of the senses.

- Name the types of sensory receptors found in the skin, and discuss their relation to the 4 cutaneous senses: touch, cold, warmth, and pain.

- Diagram the electrical responses of a sensory receptor to graded increases in stimulus strength, and describe the relationship between the size of these generator potentials and the number of impulses generated in the sensory nerve that innervates or contains the receptor.

- Define adaptation, and explain its occurrence.

- Define the doctrine of specific nerve energies and the law of projection.

- Explain how the body senses differences in the intensity of stimuli.

GENERAL QUESTIONS

1. Discuss sensory units and their recruitment as the intensity of a stimulus increases.

2. What is the mathematical relationship between the intensity of a stimulus and the intensity of the sensation that is produced?

3. What is the difference between tonic and phasic receptors? What physiologic functions are subserved by each type?

4. What part of a pacinian corpuscle produces its generator potential? Where are action potentials in the sensory nerve produced? What is the function of the multiple lamellas of connective tissue that cover the ending of the sensory nerve in a pacinian corpuscle?

5. Why do people see lightning and hear thunder rather than hear lightning and see thunder?

6. What are chemoreceptors, nociceptors, and teleceptors? Give examples of each.

7. Distinguish between a sensory receptor and a sense organ.

8. Amputees may feel pain and other sensations in a limb that is no longer there, and this can be a distressing medical problem. Explain the occurrence of pain in the absent limb. How would you treat a patient with such a "phantom limb"?

9. Discuss the possible mechanisms by which generator potentials are produced in mechanoreceptors.

MULTIPLE-CHOICE QUESTIONS

In the following questions, select the single best answer.

1. Pacinian corpuscles are—
   a. a type of temperature receptor.
   b. usually innervated by A δ nerve fibers.
   c. rapidly adapting touch receptors.
   d. slowly adapting touch receptors.
   e. pain receptors.

2. Adaptation to a sensory stimulus produces—
   a. a diminished sensation when other types of sensory stimuli are applied near a given stimulus.
   b. a more intense sensation when a given stimulus is applied repeatedly.
   c. a sensation localized to the hand when the nerves of the brachial plexus are stimulated.
   d. a diminished sensation when a given stimulus is applied steadily for a period of time.
   e. failure to perceive the stimulus when one's attention is directed to another matter.

3. What 2 words describe the relationships (1) between stimulus intensity and amplitude of the generator potential and (2) between stimulus intensity and the number of action potentials that are generated?
   a. direct, inverse.
   b. inverse, direct.
   c. inverse, inverse.
   d. direct, direct.
   e. direct, unrelated.

In the following questions, one or more than one of the answers may be correct. Select—

   A if (1), (2), and (3) are correct;
   B if (1) and (3) are correct;
   C if (2) and (4) are correct;
   D if only (4) is correct; and
   E if all are correct.

4. Which of the following sensations can be generated by impulses initiated in naked nerve endings?

(1) pain.
(2) touch.
(3) temperature.
(4) hearing.

A   B   C   D   E

5. Which of the following aspects of the external or internal environment are sensed?
(1) changes in pressure of air outside the body.
(2) plasma glucose.
(3) linear acceleration.
(4) plasma osmolality.

A   B   C   D   E

6. Which of the following determines the nature (as opposed to the intensity) of a sensation?
(1) the type of receptor.
(2) the size of the generator potential.
(3) the part of the brain where the specific neural pathways from the receptor terminate.
(4) the amplitude of the action potentials in the sensory nerves from the receptor.

A   B   C   D   E

7. A sensation felt in one hand can be produced by stimulation of—
(1) the afferent neurons from the hand as they enter the spinal cord.
(2) the spinothalamic tract neurons on which the afferent neurons from the hand terminate.
(3) the hand area of the sensory cortex on the opposite side of the body.
(4) the hand area of the sensory cortex on the same side of the body.

A   B   C   D   E

8. Impulses in sensory nerves are generated by receptors located in sense organs. Sometimes the receptors are confused with the sense organs. Which of the following receptors and sense organs are correctly paired?
(1) rods and cones : eye.
(2) hair cells : inner ear.
(3) stretch receptors : carotid sinus.
(4) hair cells : taste buds.

A   B   C   D   E

In the following questions, indicate whether the item on the left is greater than **(G)**, the same as **(S)**, or less than **(L)** the item on the right.

9. Amplitude of generator potential with weak stimulus.   G S L   Amplitude of generator potential with strong stimulus.

10.	Amplitude of action potentials in sensory nerve with weak stimulus.	**G**	**S**	**L**	Amplitude of action potentials in sensory nerve with strong stimulus.
11.	Frequency of action potentials in sensory nerve with weak stimulus.	**G**	**S**	**L**	Frequency of action potentials in sensory nerve with strong stimulus.
12.	Sensation of pain when warmth receptors are subjected to weak stimulation.	**G**	**S**	**L**	Sensation of pain when warmth receptors are subjected to strong stimulation.
13.	Rate of discharge in nerve from muscle spindle with sustained stretch.	**G**	**S**	**L**	Rate of discharge in nerve from touch receptor with sustained touch.

REFERENCES

Darian-Smith I (editor): *Sensory Processing*. Section 1, Vol 3, of: *Handbook of Physiology*. American Physiological Society, 1985.

Schmidt RF (editor): *Fundamentals of Sensory Physiology*. Chapters 1 and 2. Springer-Verlag, 1978.

6 Reflexes

This chapter describes the components that make up the reflex arc, the neural substrate for reflex responses. Monosynaptic and polysynaptic spinal reflexes are discussed, using stretch and withdrawal reflexes as examples. The general properties of reflexes are also analyzed.

OBJECTIVES

The material in the chapter should help students to—

- List the different types of nerve fibers in the dorsal and ventral roots of the spinal nerves, and outline the functional implications of the differences between them.

- Distinguish between and compare monosynaptic and polysynaptic reflexes.

- Give examples of stretch reflexes, including those that are frequently tested clinically.

- Describe the muscle spindles and analyze their function, with particular attention to how they operate as part of a feedback system to maintain muscle length.

- Describe the sensory and motor innervation of the muscle spindles and the type and function of each of the nerve fibers that make up this innervation.

- Understand the effects of variation in the rate of γ efferent discharge.

- Define reciprocal innervation, inverse stretch reflex, clonus, and lengthening reaction.

- Understand the term muscle tone, and discuss the factors that regulate muscle tone.

- Define and explain fractionation and occlusion.

GENERAL QUESTIONS

1. What is a "final common path"? Discuss its physiologic basis and importance.

2. Why do paraplegic patients sometimes urinate, defecate, and show marked withdrawal responses and fluctuations in blood pressure in response to relatively minor noxious stimuli distal to their level of cord transection?

3. What is the Bell-Magendie Law? What are its physiologic implications?

4. Define local sign, and explain the mechanism responsible for it.

5. Why does a strong noxious stimulus produce a prolonged withdrawal response?

6. Discuss the control of γ efferent discharge. Describe how variations in the rate of γ efferent discharge alter the sensitivity of muscle spindles to stretch.

7. In a normal subject, the central delay for a given reflex response was found to be 0.9 ms. Is it likely that the response is mediated by a monosynaptic or a polysynaptic pathway? Explain your answer.

8. It is frequently stated that Golgi tendon organs are part of a servomechanism that controls muscle force. Analyze and discuss this statement.

9. At what points in the reflex arc can reflex responses be graded and modified?

MULTIPLE-CHOICE QUESTIONS

In the following questions, select the single best answer.

1. The inverse stretch reflex—
   a. has a lower threshold than that of the stretch reflex.
   b. is probably a monosynaptic reflex.
   c. is probably a disynaptic reflex with a single interneuron inserted between the afferent and efferent limbs.
   d. is probably a polysynaptic reflex with many interneurons inserted between the afferent and efferent limbs.
   e. requires the discharge of central neurons that release gamma-aminobutyrate (GABA).

2. When γ motor neuron discharge increases at the same time as α motor neuron discharge to a muscle—
   a. there is prompt inhibition of discharge in spindle Ia afferents.
   b. the contraction of the muscle is prolonged.
   c. the muscle will not contract.
   d. the number of impulses in spindle Ia afferents is less than when α discharge alone is increased.
   e. the number of impulses in spindle Ia afferents is greater than when α discharge alone is increased.

3. When a normally innervated skeletal muscle is stretched, the initial response is contraction, but with increasing stretch, the muscle suddenly relaxes. This relaxation occurs because—
   a. with strong stretch, γ efferent discharge is decreased.
   b. with strong stretch, the discharge from the annulospiral endings of afferent nerve fibers is inhibited.
   c. with strong stretch, there is decreased activity in the afferent nerve fibers from the Golgi tendon organs.
   d. with strong stretch, there is increased activity in the afferent nerve fibers from the Golgi tendon organs.
   e. as a result of reciprocal innervation, there is increased discharge in the afferent nerve fibers from the antagonists to the stretched muscle.

In the following questions, one or more than one of the answers may be correct. Select—

  A if (1), (2), and (3) are correct;
  B if (1) and (3) are correct;
  C if (2) and (4) are correct;
  D if only (4) is correct; and
  E if all are correct.

4. A reflex action—
   (1) cannot be modified by impulses in other parts of the nervous system.
   (2) may involve simultaneous contraction of some muscles and relaxation of others.
   (3) may involve either somatic or visceral responses but never both simultaneously.
   (4) always involves transmission across at least one synapse.

   A   B   C   D   E

5. In a polysynaptic reflex, which of the following happen when the strength of the adequate stimulus is increased?
   (1) the latency is decreased.
   (2) the amplitude of the motor response is increased.
   (3) the motor response spreads to include other muscles and even other limbs.
   (4) there is increased inhibition of stretch reflexes.

   A   B   C   D   E

6. When all the sensory inputs that produce a polysynaptic reflex are dissected out and stimulated one after another—
   (1) the sum of the tensions developed with each stimulus will be greater than the tension developed when all of the inputs are stimulated simultaneously.
   (2) no tension is developed, because it is necessary for spatial facilitation to occur if a response is to be observed.
   (3) supramaximal stimulation of any one input does not produce as strong a contraction as that produced by direct stimulation of the muscle.
   (4) the central delay will be found to be about 0.3 ms.

   A   B   C   D   E

7. When a dorsal root in the lower thoracic region is sectioned, large numbers of the following types of nerve fibers are interrupted—
   (1) preganglionic sympathetic fibers.
   (2) afferent fibers from cutaneous receptors.
   (3) motor neurons to skeletal muscle.
   (4) afferent fibers from the abdominal viscera.

   A   B   C   D   E

8. In Fig 6–1, A, B, and C are records of action potentials in afferent nerve fibers from a muscle before and after the muscle was stretched, as shown by the change in muscle length.
   (1) A is the type of record produced in fibers from Golgi tendon organs.
   (2) B is the type of record produced in nuclear chain fibers.
   (3) C is the type of record produced in nuclear bag fibers.
   (4) A is the type of record produced in nuclear bag fibers.

   A   B   C   D   E

9. When a ventral root in the lower thoracic region is sectioned, large numbers of the following types of nerve fibers are interrupted—
   (1) preganglionic sympathetic fibers.
   (2) afferent fibers from cutaneous receptors.

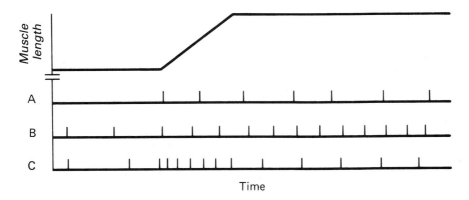

Figure 6–1.

(3) motor neurons to skeletal muscle.
(4) afferent fibers from abdominal viscera.

A   B   C   D   E

10. Decreased discharge in the Ia afferent fibers close to the annulospiral endings in a muscle spindle will be observed when—
    (1) the muscle is passively shortened.
    (2) the muscle contracts in response to stimulation of motor fibers to the extrafusal muscle fibers.
    (3) the ventral root of the spinal nerve that supplies the muscle is cut.
    (4) the intrafusal fibers contract in response to stimulation of the γ efferent fibers that innervate them.

A   B   C   D   E

In the following questions, indicate whether the item on the left is greater than (**G**), the same as (**S**), or less than (**L**) the item on the right.

11. Response of a skeletal muscle to stretch after cutting the sensory nerves that supply it.

    G   S   L

    Response of a skeletal muscle to stretch after cutting the motor nerves that supply it.

12. Response of skeletal muscle to stretch in an anxious patient.

    G   S   L

    Response of skeletal muscle to stretch in a relaxed patient.

13. During stretch reflex, rate of discharge in motor nerve to protagonist muscles.

    G   S   L

    During stretch reflex, rate of discharge in motor nerve to antagonist muscles.

14. Response of flexor muscles in a limb when a noxious stimulus is applied to the ipsilateral limb.

    G   S   L

    Response of flexor muscles in a limb when a noxious stimulus is applied to the contralateral limb.

15. Response of extensor muscles in a limb when a noxious stimulus is applied to the ipsilateral limb.

    G   S   L

    Response of extensor muscles in a limb when a noxious stimulus is applied to the contralateral limb.

16. Number of neurons in-
volved in stretch reflex.

   G S L   Number of neurons involved in withdrawal
reflex.

17. Central inhibitory state 1
day after transection of the
spinal cord in humans.

   G S L   Central inhibitory state 1
month after transection of
the spinal cord in humans.

18. Rhythmic contractile response to sustained stretch
of a muscle when rate of γ
efferent discharge is high.

   G S L   Rhythmic contractile response to sustained stretch
of a muscle when rate of γ
efferent discharge is low.

19. Resistance to stretch in a
flaccid muscle.

   G S L   Resistance to stretch in a
spastic muscle.

20. Latency of a monosynaptic
reflex response.

   G S L   Latency of a polysynaptic
reflex response.

REFERENCES

Bronisch KW: *The Clinically Important Reflexes*. Grune & Stratton, 1952.
Granit R, Pompeiano O (editors): Reflex control of posture and movement. *Prog Brain Res* 1979;**50**:1.
Sherrington CS: *The Integrative Action of the Nervous System*. Cambridge Univ Press, 1947.

Cutaneous, Deep, & Visceral Sensation

7

Chapter 7 describes the sensory pathways that mediate touch, proprioception, warmth, cold, pain, itching, and sensations produced by combinations of stimuli. Pain is considered in detail, including the characteristics of so-called fast pain and slow pain, the emotional concomitants of pain, deep pain, and pain from visceral structures.

OBJECTIVES

The material in the chapter should help students to—

- Outline the neural connections and pathways from the skin, deep tissues, and viscera to the cerebral cortex that mediate the various specific sensations.

- Describe the substantia gelatinosa, and discuss its physiologic significance.

- Describe the areas in which the sensory projection neurons from the thalamus terminate in the sensory cortex.

- Distinguish between touch and proprioception in terms of the receptors and pathways that mediate each sensation.

- Name the types of nerve fibers that mediate warmth and cold in peripheral nerves, and describe where impulses generated in warmth and cold receptors terminate in the cortex.

- Name the receptors that mediate pain, and explain the differences between fast and slow pain.

- List the various neurosurgical procedures that have been employed for relief of intractable pain, and explain how they produce relief.

- Compare superficial, deep, and visceral pain.

- Define hyperalgesia.

- Describe and explain referred pain.

- List and explain ways that the sensation of pain can be inhibited.

- Explain vibratory sensibility, 2-point discrimination, and stereognosis.

GENERAL QUESTIONS

1. What are the adequate stimuli for warmth and cold?

2. What is the adequate stimulus for pain? Are pain receptors specific? What chemical agents may initiate impulses in pain fibers?

3. What is the thalamic syndrome? Explain its pathophysiology.

4. Why do tumors that arise in the center of the upper spinal cord cause loss of pain and temperature sensations first in the upper segments of the body and only later in the lower segments? What would you expect to be the corresponding order of loss of proprioception?

5. Describe the location of the 2 touch pathways in the spinal cord. What type of touch information does each conduct?

6. Analyze the rationale for and the advantages and disadvantages of the use of prefrontal lobotomy in the treatment of intractable pain in cancer patients.

7. Discuss P factor and the role it plays in muscle pain.

8. Why is visceral pain poorly localized? Why is it particularly unpleasant?

9. What are opioid peptides? Describe how they alter pain sensation.

MULTIPLE-CHOICE QUESTIONS

In the following questions, select the single best answer.

1. An anterolateral cordotomy is performed that produces relief of pain in the right leg. It is effective because it interrupts the—
   a. left dorsal column.
   b. left ventral spinothalamic tract.
   c. right lateral spinothalamic tract.
   d. left lateral spinothalamic tract.
   e. right corticospinal tract.

2. Stimulation of which of the following might be expected to produce itching?
   a. dorsal root C fibers.
   b. B fibers in peripheral nerves.
   c. dorsal columns of the spinal cord.
   d. touch receptors.
   e. motor fibers in the ventral roots of the spinal nerves.

3. Visceral pain—
   a. shows relatively rapid adaptation.
   b. is mediated by B fibers in the dorsal roots of the spinal nerves.
   c. can sometimes be relieved by applying an irritant to the skin.
   d. is similar to "fast pain" produced by noxious stimulation of touch receptors.
   e. can be produced by marked and prolonged stimulation of touch receptors.

4. The distance by which 2 touch stimuli must be separated to be perceived as 2 separate stimuli is greatest on—
   a. the lips.
   b. the palm of the hand.
   c. the back of the scapula.
   d. the dorsum of the hand.
   e. the tips of the fingers.

5. Injecting a local anesthetic into an area to which visceral pain is referred sometimes abolishes the pain. When this occurs, it indicates that—
   a. the sensory neurons involved send one branch to the viscus and another to the area to which the pain is referred.

b. the referral is psychosomatic.

c. the individual with the pain has previously had a painful injury in the area to which the pain is referred.

d. impulses in afferents from the viscus somehow facilitate transmission in the sensory pathway from the area to which the pain is referred.

e. the afferents from the viscus and those from the area to which the pain is referred end on the same second-order sensory neurons.

In the following questions, select—
    A if the item is associated with **(a)** below,
    B if the item is associated with **(b)** below,
    C if the item is associated with both **(a)** and **(b)**, and
    D if the item is associated with neither **(a)** nor **(b)**.

(a) dorsal root C fibers.
(b) A fibers.

6. Cold   **A   B   C   D**

7. Warmth   **A   B   C   D**

8. Pain   **A   B   C   D**

9. Touch   **A   B   C   D**

10. Muscle stretch   **A   B   C   D**

In the following questions, one or more than one of the answers may be correct. Select—
    A if (1), (2), and (3) are correct;
    B if (1) and (3) are correct;
    C if (2) and (4) are correct;
    D if only (4) is correct; and
    E if all are correct.

11. Which of the following need to be intact for normal stereognosis?
 (1) dorsal columns.
 (2) parietal lobe.
 (3) touch pathways.
 (4) lateral spinothalamic tract.

     A   B   C   D   E

12. In which of the following tracts in the spinal cord do second-order sensory neurons with cell bodies in the dorsal horn ascend to more rostral spinal segments or to the brain?
 (1) ventral corticospinal tract.
 (2) lateral spinothalamic tract.
 (3) anterior vestibulospinal tract.
 (4) ventral spinothalamic tract.

     A   B   C   D   E

13. Pain due to partial obstruction of the ureter may be felt in the testicle because—
    (1) nerve fibers from the testicles and nerve fibers from the ureter may end on the same neurons in the spinal cord.
    (2) embryologically, the testicles arise from the same dermatome as the ureter.
    (3) nerve fibers from the ureter may send collaterals to neurons innervated by afferent fibers from the testicles.
    (4) pain causes a general increase in the excitability of neurons in the spinal cord.

    A   B   C   D   E

14. Primary afferent neurons terminate in the following regions—
    (1) thalamus.
    (2) superficial layers of the ventral horn.
    (3) lateral portion of the contralateral intermediolateral gray column of the spinal cord.
    (4) substantia gelatinosa.

    A   B   C   D   E

15. Which of the following may be chemical mediators of pain, ie, substances that are produced locally and act on naked nerve endings to initiate impulses in pain fibers?
    (1) histamine.
    (2) prostaglandins.
    (3) kinins.
    (4) adenosine.

    A   B   C   D   E

In the following questions, indicate whether the item on the left is greater than **(G)**, the same as **(S)**, or less than **(L)** the item on the right.

16. Vibration sensitivity in a patient with pernicious anemia. **G S L** Vibration sensitivity in a patient with iron deficiency anemia.

17. β-Endorphin level in cerebrospinal fluid of a patient receiving morphine. **G S L** β-Endorphin level in cerebrospinal fluid of a patient being treated with acupuncture.

18. Number of impulses reaching sensory cortex in a patient being treated with morphine. **G S L** Number of impulses reaching sensory cortex in a patient being treated with naloxone, a morphine antagonist.

19. Number of impulses in primary afferent neurons in a patient being treated with morphine. **G S L** Number of impulses in primary afferent neurons in a patient being treated with naloxone.

20. Rate of conduction in afferent neurons from cold receptors.   **G S L**   Rate of conduction in afferent neurons from pressure receptors.

21. Loss of proprioceptive sensation in a patient with disease of the dorsal columns.   **G S L**   Loss of warmth sensation in the same patient.

22. Deficit in processing of sensory information in sensory area I (SI) after lesion in SII.   **G S L**   Deficit in processing of sensory information in sensory area II (SII) after lesion in SI.

23. Effects of parietal lobe lesions on stereognosis.   **G S L**   Effect of frontal lobe lesions on stereognosis.

REFERENCES

Kandel ER, Schwartz JH, (editors): *Principles of Neural Science,* 2nd ed. Chapters 16–18. Elsevier, 1985.

Schmidt RF (editor): *Fundamentals of Sensory Physiology.* Chapter 3: Somatovisceral sensibility. Springer-Verlag, 1978.

Vision

Chapter 8 considers the functional anatomy of the visual system, then analyzes in order the processes involved in vision: formation of the visual image on the retina, conversion of light energy to electrical responses in the rods and cones, processing of impulses in the retina, transmission of impulses via the lateral geniculate bodies to the visual cortex, and the events that occur in the visual cortex. Color vision and the control of eye movements are also reviewed.

OBJECTIVES

The material in the chapter should help students to—

● Describe the various parts of the eye, and list the functions of each.

● Trace the neural pathways that transmit visual information from the rods and cones to the visual cortex.

● Explain how light rays in the environment are brought to a focus on the retina and the role of accommodation in this process.

● Define hyperopia, myopia, astigmatism, presbyopia, and strabismus.

● Describe the electrical responses produced by rods and cones, and explain how these responses are produced.

● Describe the electrical responses seen in bipolar cells, horizontal cells, amacrine cells, and ganglion cells, and comment on the function of each type of cell.

● Describe the responses of simple and complex cells in the visual cortex and the functional organization of the areas of the cerebral cortex that process visual information.

● Define and explain dark adaptation and visual acuity.

● Describe the functions of cones, the neural pathways to the visual cortex, and the visual cortex in color vision.

● Name the 4 types of eye movements and the function of each.

GENERAL QUESTIONS

1. What type of visual field defect is produced by each of the following lesions and why?
   a. a lesion of one optic nerve.
   b. a lesion of one optic tract.
   c. a lesion of the optic chiasm.
   d. lesions in various parts of the geniculocalcarine tract.
   e. bilateral lesions destroying the visual cortex (area 17).

2. What is the near point of vision? Why does it recede throughout life?

3. What is an Argyll Robertson pupil? Explain why it occurs in various diseases of the nervous system.

4. How many different kinds of photosensitive pigments are found in the human retina? Discuss their chemistry and how they produce electrical responses in the retina.

5. What are "on center" and "off center" cells? Define lateral inhibition, and comment on its general physiologic significance.

6. Some drugs that inhibit monoamine oxidase have been reported to impair color vision. Explain how such impairment might be produced.

7. What are orientation columns and ocular dominance columns? How are they mapped experimentally? What are their functions?

8. Discuss parallel processing of visual information in the cerebral cortex.

9. What is binocular vision? Why is it limited to part of the visual field? Discuss its role and the role of monocular visual processes in depth perception.

10. How is color blindness diagnosed? What are the various types of color blindness? Why does inherited color blindness characteristically skip generations and occur in males rather than in females?

11. What is amblyopia ex anopsia? Explain its pathogenesis.

12. Adaptation of what cells is responsible for the upper left portion of the curve in Fig 8–1? Adaptation of what cells is responsible for the lower right portion? In each instance, explain why the curve levels off.

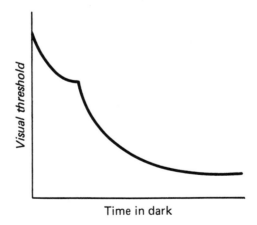

Time in dark

Figure 8–1. Dark adaptation.

MULTIPLE-CHOICE QUESTIONS

In the following questions, select the single best answer.

1. If the principal focal distance of a lens is 0.75 m, its refractive power is—
   a. 0.25 diopter.
   b. 0.75 diopter.
   c. 1.0 diopter.

    d. 1.33 diopters.
    e. 10.3 diopters.

2. Visual accommodation involves—
    a. increased tension on the lens ligaments.
    b. a decrease in the curvature of the lens.
    c. relaxation of the sphincter muscle of the iris.
    d. contraction of the ciliary muscle.
    e. increased intraocular pressure.

3. When a visual stimulus falls on a given point in the retina for a long time—
    a. the image becomes more clearly focused.
    b. there is adaptation in the superior colliculus.
    c. the discharge rate in the bipolar cells increases.
    d. the pupils constrict.
    e. the image fades and disappears.

4. The fovea of the eye—
    a. has the lowest light threshold.
    b. is the region of highest visual acuity.
    c. contains only red and green cones.
    d. contains only rods.
    e. is situated over the head of the optic nerve.

5. Which of the following parts of the eye has the greatest concentration of rods?
    a. ciliary body.
    b. iris.
    c. optic disk.
    d. fovea.
    e. parafoveal region.

6. The following events that occur in rods in response to light are listed in random sequence:
    1. activation of transducin.
    2. decreased release of synaptic transmitter.
    3. structural changes in rhodopsin.
    4. closure of Na^+ channels.
    5. decrease in intracellular cyclic GMP.

What is the sequence in which they normally occur?

    a. 2, 1, 3, 5, 4.
    b. 1, 2, 3, 5, 4.
    c. 5, 3, 1, 4, 2.
    d. 3, 1, 5, 4, 2.
    e. 3, 1, 4, 5, 2.

In the following questions, one or more than one of the answers may be correct. Select—
    A if (1), (2), and (3) are correct;
    B if (1) and (3) are correct;
    C if (2) and (4) are correct;
    D if only (4) is correct; and
    E if all are correct.

7. Which of the following affect visual acuity?
    (1) contrast between stimulus and background.
    (2) cataracts.

(3) incident light.
(4) astigmatism.

A   B   C   D   E

8. Vitamin A is a precursor for the synthesis of—
(1) somatostatin.
(2) cone pigments.
(3) the pigment of the iris.
(4) rhodopsin.

A   B   C   D   E

9. In which of the following does transduction of the light signal involve a G protein?
(1) red cones.
(2) green cones.
(3) blue cones.
(4) rods.

A   B   C   D   E

10. The innervation of the eye and its muscles includes—
(1) afferent fibers mediating vision.
(2) somatic efferent fibers to the medial and lateral rectus muscles.
(3) visceral efferent fibers to the ciliary muscle.
(4) visceral efferent fibers to the iris muscle.

A   B   C   D   E

11. The visual receptors connect *directly* to the ganglion cells in the retina via—
(1) horizontal cells.
(2) amacrine cells.
(3) cones.
(4) bipolar cells.

A   B   C   D   E

12. Which of the following parts of the brain are concerned with processing visual information?
(1) medial geniculate body.
(2) lateral portion of temporal lobe.
(3) postcentral gyrus.
(4) primary visual cortex (area 17).

A   B   C   D   E

In the following questions, indicate whether the item on the left is greater than **(G)**, the same as **(S)**, or less than **(L)** the item on the right.

13. Depression of eyeball (pupil moves down) produced by contraction of superior oblique muscle.   **G S L**   Depression of eyeball produced by contraction of inferior oblique muscle.

14. Color vision subserved by rods. **G S L** Color vision subserved by cones.

15. Change in curvature of anterior surface of lens during accommodation. **G S L** Change in curvature of posterior surface of lens during accommodation.

16. Membrane potential of synaptic end of visual receptor in light. **G S L** Membrane potential of synaptic end of visual receptor in darkness.

17. Number of action potentials in right visual cortex after stimulation of medial half of right retina. **G S L** Number of action potentials in left visual cortex after stimulation of medial half of right retina.

18. Loss of vision after lesion of left medial geniculate body. **G S L** Loss of vision after lesion of left lateral geniculate body.

19. Visual acuity in a person with 20/15 vision. **G S L** Visual acuity in a person with 15/20 vision.

20. Principal focal distance of lens used to correct defect in a patient with severe hyperopia. **G S L** Principal focal distance of lens used to correct defect in a patient with mild hyperopia.

21. Response to colored stimuli of cells in visual cortex that stain for cytochrome oxidase. **G S L** Response to colored stimuli of cells in visual cortex that do not stain for cytochrome oxidase.

REFERENCES

Hubel DH, Weisel TN: Brain mechanism of vision. *Sci Am* (Sept) 1979;**241**:150.
Schmidt RF (editor): *Fundamentals of Sensory Physiology*. Chapter 4: Vision. Springer-Verlag, 1978.

Hearing & Equilibrium

9

Chapter 9 starts with a brief review of the functional anatomy of the external, middle, and inner ear. The physiology of hair cells is reviewed, and the chapter discusses how sound waves are converted into auditory sensations and reviews the neural pathways from the cochlea via the auditory nerve and the brain stem to the auditory cortex. The sense of orientation of the body in space is also considered, including detection of rotational acceleration by receptors in the semicircular canals, detection of linear acceleration by receptors in the saccule and utricle, and relay of information via vestibular pathways to all parts of the brain stem.

OBJECTIVES

The material in the chapter should help students to—

- Describe the way that movements of molecules in the air are converted into movements of the foot plate of the stapes.

- Outline the properties of traveling waves, and describe how, via these waves, particular movements of the foot plate of the stapes produce maximal deformations of the basilar membrane at particular points.

- Describe how deformations in the basilar membrane are converted to impulses in auditory nerve fibers.

- Trace the path of auditory impulses in the neural pathways from the cochlea to the auditory cortex.

- Discuss the structure and function of the auditory cortex.

- Explain how pitch and loudness are coded in the auditory pathways.

- Discuss the mechanisms that permit the origin of a sound to be determined.

- Describe the various forms of deafness.

- Explain how the receptors in the semicircular canals detect rotational acceleration and the receptors in the saccule and utricle detect linear acceleration.

- Describe and explain the effects of labyrinthectomy.

- List the major sensory inputs that provide the information which is synthesized in the brain into the sense of position in space.

GENERAL QUESTIONS

1. Otosclerosis is a disease in which the foot plate of the stapes becomes rigidly attached to the oval window. Why does this cause deafness?

2. What is the function of the tectorial membrane in the cochlea?

3. What is the function of the otoliths?

4. What sound frequencies are audible to humans? At what sound frequency is the auditory threshold the lowest?

5. Define and explain masking.

6. What is the tympanic reflex, and what is its function?

7. Compare ossicular, air, and bone conduction. What is a fenestration operation, and why does it improve hearing in patients with conduction deafness?

8. Describe how a tuning fork can be used to distinguish between nerve deafness and conduction deafness.

9. What is the vestibulo-ocular reflex, and how does it relate to nystagmus?

10. If the vision of individuals lacking vestibular function is obscured while they are diving in water, they may swim away from rather than toward the surface and drown. Why does this occur?

MULTIPLE-CHOICE QUESTIONS

In the following questions, select the single best answer.

1. Sound intensity is measured in—
   a. diopters.
   b. daltons.
   c. torrs.
   d. decibels.
   e. pounds.

2. In humans, the primary auditory cortex is located in the—
   a. limbic system.
   b. posterior part of the occipital lobe.
   c. posterior part of the parietal lobe.
   d. postcentral gyrus.
   e. superior part of the temporal lobe.

3. Postrotatory nystagmus is caused by—
   a. continued movement of perilymph over hair cells that have their processes embedded in the tectorial membrane.
   b. continued movement of aqueous humor over the ciliary body in the eye.
   c. continued movement of cerebrospinal fluid over the parts of the brain stem that contain the vestibular nuclei.
   d. continued movement of endolymph in the semicircular canals, with consequent bending of the cupula and stimulation of hair cells.
   e. a disease process affecting the vestibular system.

4. The basilar membrane of the cochlea—
   a. is unaffected by movement of fluid in the scala vestibuli.
   b. covers the oval window and the round window.
   c. vibrates in a pattern determined by the form of the traveling wave in the fluids in the cochlea.
   d. is under tension.
   e. vibrates when the body is subjected to linear acceleration.

5. Which of the following are *incorrectly* paired?
    a. timbre (quality) of sound : harmonic vibrations.
    b. pitch of sound : frequency of sound wave.
    c. pitch of sound : point of greatest displacement of the basilar membrane.
    d. direction of sound : frequency of impulses in an auditory nerve fiber.
    e. loudness of sound : amplitude of sound wave.

In the following questions, select—

>   **A** if the item is associated with (a) below,
>   **B** if the item is associated with (b) below,
>   **C** if the item is associated with both (a) and (b), and
>   **D** if the item is associated with neither (a) nor (b).

>   **(a)** Sound transmission.
>   **(b)** Linear acceleration.

6.	Structure V in Fig 9–1	A	B	C	D	E
7.	Structure W in Fig 9–1	A	B	C	D	E
8.	Structure X in Fig 9–1	A	B	C	D	E
9.	Structure Y in Fig 9–1	A	B	C	D	E
10.	Structure Z in Fig 9–1	A	B	C	D	E

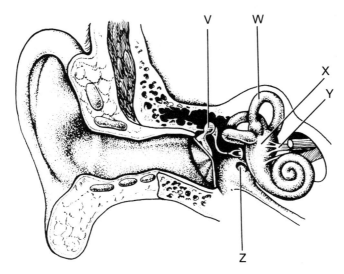

Figure 9–1. Human ear.

In the following questions, one or more than one of the answers may be correct.
Select—

>   **A** if (1), (2), and (3) are correct;
>   **B** if (1) and (3) are correct;
>   **C** if (2) and (4) are correct;
>   **D** if only (4) is correct; and
>   **E** if all are correct.

11. In the ear, hair cells—
    (1) are innervated by the olivocochlear bundle.
    (2) are found in the maculas of the saccule and utricle.
    (3) are found in the cristae of the semicircular canals.
    (4) line the auditory (eustachian) tube.

    A   B   C   D   E

12. Sound localization—
    (1) depends in part on the difference between the quality of a sound coming from in front of an individual and that of a sound coming from behind the individual.
    (2) is disrupted by lesions of the auditory cortex.
    (3) depends in part on the more rapid arrival of sound at the ear on the side nearest the sound.
    (4) depends in part on neurons that only respond maximally when impulses reaching them from the 2 ears are separated by a fixed interval.

    A   B   C   D   E

13. Which of the following are *correctly* paired?
    (1) tympanic membrane : manubrium of malleus.
    (2) helicotrema : apex of cochlea.
    (3) foot plate of stapes : oval window.
    (4) otoliths : semicircular canals.

    A   B   C   D   E

14. The direction of nystagmus is vertical when a subject is rotated—
    (1) after warm water is put in one ear.
    (2) with the head tipped backward.
    (3) after cold water is put in both ears.
    (4) with the head tipped sideways.

    A   B   C   D   E

15. Which of the following contribute to the sense of orientation in space?
    (1) visual cues.
    (2) input from touch and pressure receptors.
    (3) input from receptors in joints.
    (4) input from the cochlea.

    A   B   C   D   E

In the following questions, indicate whether the item on the left is greater than (**G**), the same as (**S**), or less than (**L**) the item on the right.

16. K^+ concentration in peri-lymph.   **G   S   L** K^+ concentration in endo-lymph.

17. K^+ concentration in peri-lymph.   **G   S   L** K^+ concentration in blood plasma.

18. Displacement of basilar membrane near apex of cochlea during a high-pitched sound.

    G   S   L

    Displacement of basilar membrane near apex of cochlea during a low-pitched sound.

19. Frequency of a low-pitched sound.

    G   S   L

    Frequency of a high-pitched sound.

20. Dizziness and related symptoms after destruction of the labyrinth on one side of the head.

    G   S   L

    Dizziness and related symptoms after destruction of the labyrinths on both sides of the head.

21. Sound threshold for bone conduction in conduction deafness.

    G   S   L

    Sound threshold for bone conduction in nerve deafness.

22. Sound threshold when the tensor tympani is contracted.

    G   S   L

    Sound threshold when the tensor tympani is relaxed.

23. At the start of rotation, movement of cupulas in semicircular canals in the direction of rotation.

    G   S   L

    At the end of rotation, movement of cupulas in semicircular canals in the direction of rotation.

24. Maximum sound frequency audible to humans.

    G   S   L

    Maximum sound frequency audible to dogs.

25. Change in membrane potential when stereocilia of a saccular hair cell are pushed toward the kinocilium.

    G   S   L

    Change in membrane potential when stereocilia of a saccular hair cell are pushed at a right angle to the kinocilium.

REFERENCES

Hudspeth AJ: The hair cells of the inner ear. *Sci Am* (Jan) 1983;**248**:54.
Pickels JO: *An Introduction to the Physiology of Hearing*. Academic Press, 1982.

Chapter 10 is concerned with olfactory pathways to the brain, how odoriferous molecules reaching the olfactory mucous membrane set up impulses in olfactory pathways, and how these impulses create the sensation of smell. The basic pathways by which impulses generated in the taste buds reach the sensory cortex are also reviewed. The mechanisms that generate impulses in taste buds are considered, and the characteristics of the substances that produce the 4 basic tastes and their interactions are analyzed.

OBJECTIVES

The material in the chapter should help students to—

- Describe the olfactory receptors and the way impulses are initiated in them.

- Outline the pathways by which impulses generated in the olfactory mucous membrane reach the cerebral cortex.

- Describe and analyze olfactory sensitivity, discrimination, and adaptation.

- Describe the taste buds, and discuss their characteristics and distribution in relation to the 4 basic tastes.

- Outline the taste pathways.

- List the substances that produce sweet, sour, bitter, and salt tastes, and comment on their interactions.

- List the factors that combine to produce flavor.

GENERAL QUESTIONS

1. Compare and contrast smell and taste. How do they interact?

2. What is the limbic system?

3. Why does sniffing improve olfactory sensitivity and discrimination?

4. Discuss the relationship between olfaction and sexual behavior.

5. What part do pain fibers in the trigeminal nerves play in olfaction?

6. Name the chemoreceptors in the body.

7. A number of seemingly unrelated substances other than sugars taste sweet. List some of these compounds, and comment on their use in the treatment of various conditions and diseases.

8. Discuss taste as a motivation for aversive learning.

MULTIPLE-CHOICE QUESTIONS

In the following questions, select the single best answer.

1. Impulses generated by olfactory receptors in the nasal mucous membrane—
   a. are relayed to the thalamus.
   b. pass through the internal capsule.
   c. are relayed to the olfactory cortex via the hypothalamus.
   d. pass to the mitral cells, and from there directly to the olfactory cortex.
   e. pass to the mitral cells, and from there to the olfactory cortex via the taste area.

2. Impulses generated in the taste buds of the tongue reach the cerebral cortex via the—
   a. thalamus.
   b. internal capsule.
   c. dorsal roots of the first cervical spinal nerves.
   d. trochlear nerve.
   e. hypoglossal nerve.

3. Impulses generated in touch receptors in the tongue reach the cerebral cortex via the—
   a. thalamus.
   b. internal capsule.
   c. dorsal roots of the first cervical spinal nerves.
   d. trochlear nerve.
   e. hypoglossal nerve.

4. The number of different odors that can be distinguished by humans is—
   a. less than 50.
   b. about 100.
   c. 200–500.
   d. 900–1000.
   e. 2000–4000.

In the following questions, one or more than one of the answers may be correct. Select—

> **A** if (1), (2), and (3) are correct;
> **B** if (1) and (3) are correct;
> **C** if (2) and (4) are correct;
> **D** if only (4) is correct; and
> **E** if all are correct.

5. Neurons in the taste pathway have cell bodies in—
   (1) ganglia on cranial nerves.
   (2) the medial geniculate body.
   (3) the nucleus of the tractus solitarius.
   (4) the hypothalamus.

   A   B   C   D   E

6. In humans, taste buds are found on—
   (1) the palate.
   (2) the pharyngeal wall.
   (3) the epiglottis.
   (4) the back of the tongue.

   A   B   C   D   E

7. Smell and taste are similar in that—
    (1) the central pathways for both relay in the same part of the thalamus.
    (2) the receptors for both are chemoreceptors.
    (3) the receptors for both are teleceptors.
    (4) both play an important role in determining the flavor of foods.

    A   B   C   D   E

8. The functions of the limbic system include—
    (1) integration of olfactory reflexes.
    (2) regulation of sexual behavior, particularly in male animals.
    (3) regulation of the balance between anger and placidity.
    (4) taste discrimination.

    A   B   C   D   E

9. Which of the following appear to act via a G protein and cyclic AMP in their receptors to generate sensations?
    (1) acid taste.   **(3)** bitter taste.
    (2) sweet taste.   **(4)** olfaction.

    A   B   C   D   E

In the following questions, indicate whether the item on the left is greater than **(G)**, the same as **(S)**, or less than **(L)** the item on the right.

10. Intensity of an odor after 2 seconds of exposure. | G S L | Intensity of an odor after 2 minutes of exposure.

11. Taste threshold concentration of saccharin (in moles). | G S L | Taste threshold of glucose (in moles).

12. Sour taste when a weak acid is placed on the center of the tongue. | G S L | Sour taste when a weak acid is placed on the edge of the tongue.

13. Percentage change in concentration of an odor-producing substance necessary to produce a detectable change in intensity of odor. | G S L | Percentage change in illumination necessary to produce a detectable change in light intensity.

14. Percentage change in concentration of a taste-producing substance necessary to produce a detectable change in intensity of taste. | G S L | Percentage change in illumination necessary to produce a detectable change in light intensity.

15. Olfactory sensitivity in a 20-year-old woman. | G S L | Olfactory sensitivity in a 70-year-old woman.

REFERENCES

Schmidt RF (editor): *Fundamentals of Sensory Physiology*. Chapter 7: Taste; and Chapter 8: Olfaction. Springer-Verlag, 1978.

Arousal Mechanisms, Sleep, & the Electrical Activity of the Brain

11

Chapter 11 deals with alertness and sleep, the mechanisms that produce these states, and the correlation between them and the electrical activity of the brain. The nonspecific sensory system in the reticular activating system is discussed. The genesis of the electroencephalogram (EEG) is also considered in the context of evoked potentials and electrical interplay between dendrites and cell bodies in the cerebral cortex.

OBJECTIVES

The material in the chapter should help students to—

- Describe the anatomy and functions of the reticular formation.

- Outline the major divisions of the thalamus and their projections to the cerebral cortex.

- Define and explain primary evoked potentials and diffuse secondary responses.

- Describe the primary types of rhythms that make up the EEG and the behavioral states that correlate with each.

- Define and explain synchronization and α block.

- Summarize the behavioral and electroencephalographic characteristics of each of the stages of slow-wave sleep.

- Summarize the electroencephalographic and other characteristics of rapid eye movement (REM) sleep, and describe the mechanisms responsible for its production.

- Describe the pattern of normal nighttime sleep in adults and the variations in this pattern from birth to old age.

- Outline the clinical uses of the EEG.

GENERAL QUESTIONS

1. What are the main electrical events in the cortex that produce the EEG?

2. Discuss general anesthesia. How is it produced? What happens to transmission in specific and nonspecific pathways during anesthesia?

3. Define somnambulism and narcolepsy, and discuss their relationship to normal sleep.

4. What causes synchrony in the EEG? What causes desynchronization? Describe the mechanisms that produce these changes, and describe the behavioral correlates of each.

5. Compare REM and non-REM sleep.

6. What is the difference between the specific sensory relay nuclei and the nonspecific projection nuclei of the thalamus?

7. Compare and contrast the factors that alter the primary evoked potential with those that alter the diffuse secondary response when a sensory organ is stimulated.

8. What is α block, and what is its physiologic significance?

9. Discuss the relation of drugs and neurotransmitters to sleep.

10. What is recurrent collateral inhibition? What is its physiologic significance?

MULTIPLE-CHOICE QUESTIONS

In the following questions, select the single best answer.

1. In a healthy, alert adult sitting with eyes closed, the dominant electro-encephalographic rhythm observed with electrodes over the occipital lobes is—
   a. delta (0.5–4 Hz).
   b. theta (4–7 Hz).
   c. alpha (8–13 Hz).
   d. beta (18–30 Hz).
   e. fast, irregular low-voltage activity.

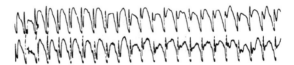

Figure 11–1.

2. The EEG record in Fig 11–1 is characteristic of—
   a. deep sleep.
   b. a learning response produced by a painful stimulus.
   c. REM sleep.
   d. a psychomotor seizure.
   e. petit mal epilepsy.

3. The electrical records in Fig 11–2 are characteristic of—
   a. deep sleep.
   b. grand mal epilepsy.
   c. REM sleep.
   d. psychomotor seizures.
   e. petit mal epilepsy.

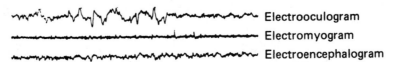
Electrooculogram
Electromyogram
Electroencephalogram

Figure 11–2. Reproduced, with permission, from Kales A et al: *Ann Intern Med* 1968;**68**:1078.

4. High-frequency stimulation of which of the following does *not* produce desynchronization?
   a. the sciatic nerve.
   b. the lateral spinothalamic tract.
   c. the midbrain reticular formation.
   d. the medial lemniscus above the midbrain.
   e. the intralaminar nuclei of the thalamus.

In the following questions, select—

        A if the item is associated with **(a)** below,
        B if the item is associated with **(b)** below,
        C if the item is associated with both **(a)** and **(b)**, and
        D if the item is associated with neither **(a)** nor **(b)**.

(a) Slow-wave sleep.
(b) REM sleep.

5. Dreaming   **A   B   C   D**

6. Sleep spindles   **A   B   C   D**

7. Sleepwalking   **A   B   C   D**

8. Bruxism   **A   B   C   D**

9. High threshold for arousal   **A   B   C   D**

In the following questions, one or more than one of the answers may be correct. Select—

        A if (1), (2), and (3) are correct;
        B if (1) and (3) are correct;
        C if (2) and (4) are correct;
        D if only (4) is correct; and
        E if all are correct.

10. Dreaming sleep—
    (1) is associated with large slow waves in the EEG.
    (2) is associated with increased muscle tone.
    (3) does not occur in old age.
    (4) is associated with ponto-geniculo-occipital (PGO) spikes.

        A   B   C   D   E

11. Neurons in the brain stem reticular formation—
    (1) can be activated by stimulation of parts of the cerebral cortex.
    (2) can be activated by stimulation of pain fibers.
    (3) transmit impulses that bypass the thalamus and project directly to the cerebral cortex.
    (4) can be inhibited by stimulation of the auditory nerve.

        A   B   C   D   E

12. The following would tend to produce slow-wave sleep—
    (1) stimulation of the basal forebrain in the region of the diagonal band.
    (2) a decrease in the number of impulses reaching the ascending reticular system.
    (3) injection of ritanserin, a drug that blocks $5HT_2$ receptors.
    (4) stimulation of the sciatic nerve.

    A   B   C   D   E

13. Stimulation of the following brain areas can produce sleep—
    (1) the posterior hypothalamus.
    (2) the occipital cortex.
    (3) the medullary reticular formation near the nucleus of the tractus solitarius.
    (4) the gracile and cuneate nuclei.

    A   B   C   D   E

14. Regular rhythmic fluctuations in electrical activity are observed in which of the following parts of the brain?
    (1) the mediobasal portion of the hypothalamus.
    (2) the cerebellar cortex.
    (3) the midbrain reticular formation.
    (4) the hippocampus.

    A   B   C   D   E

In the following questions, indicate whether the item on the left is greater than **(G)**, the same as **(S)**, or less than **(L)** the item on the right.

15. Frequency of EEG waves in a subject in coma.   **G S L**   Frequency of EEG waves in a subject in REM sleep.

16. Amplitude of EEG waves over a subdural hematoma (a collection of blood over the cerebral cortex).   **G S L**   Amplitude of EEG waves over an epileptic focus in the cortex.

17. Frequency of EEG waves in deep non-REM sleep.   **G S L**   Frequency of EEG waves in REM sleep.

18. Frequency of episodes of REM sleep after first falling asleep.   **G S L**   Frequency of episodes of REM sleep after 6 hours of sleep.

19. Area of cortex from which the diffuse secondary response can be recorded.   **G S L**   Area of cortex from which the primary evoked potential can be recorded.

20. Degree of synchrony during slow-wave sleep.   **G S L**   Degree of synchrony during alert attention.

| 21. | Number of different types of sensory input producing responses in a thalamo-cortical neuron in the ventrobasal nuclei of the thalamus. | **G** | **S** | **L** | Number of different types of sensory input producing responses in a thalamocortical neuron in the midline nuclei of the thalamus. |

REFERENCES

deGroot J, Chusid JG: *Correlative Neuroanatomy,* 20th ed. Appleton & Lange, 1988.

Shepherd GM: *Neurobiology,* 2nd ed. Chapter 25: Biorhythms. Oxford Univ Press, 1988.

12 Control of Posture & Movement

Chapter 12 is a review of somatic motor function. It points out that voluntary motor activity is initiated in the association areas of the cerebral cortex, programmed in the basal ganglia and cerebellum, and brought about in large part via the corticospinal and corticobulbar tracts as a final common path from the motor cortex. Other pathways also contribute, but primarily they provide the postural adjustments and background for movement. In addition to its planning function, the cerebellum provides the feedback control that makes movement smooth and coordinated. The functions of these components are reviewed, including the hierarchical organization of postural reflexes at the spinal, medullary, midbrain, and cortical levels. In addition, the abnormalities produced by diseases of the basal ganglia and cerebellum in humans are described and analyzed in terms of the insights they provide into the normal functions of these parts of the brain.

OBJECTIVES

The material in the chapter should help students to—

- Describe in general terms how posture and movement are regulated, and outline the function of each of the main components of the regulatory systems.

- Describe the cortical motor areas, and the corticospinal and corticobulbar tracts.

- Discuss the function of the corticospinal and corticobulbar tracts in relation to skilled voluntary movement.

- Define the concept of release, and describe the hierarchical organization of motor function.

- Define spinal shock, and explain the initial and long-term changes in reflexes that follow transection of the spinal cord.

- List and explain the long-term complications that make it difficult to treat patients with spinal cord injuries.

- Define decerebrate and decorticate rigidity, and comment on the cause of each.

- Describe the postural reflexes that are integrated in the medulla oblongata, the pons, the midbrain, and the cerebral cortex.

- Describe the basal ganglia, and list the pathways that interconnect them, along with the neurotransmitters in each pathway.

- Describe and explain the symptoms of Parkinson's disease.

- List the pathways to and from the cerebellum and the connections of each within the cerebellum.

- Discuss the functions of the cerebellum and the neurologic abnormalities produced by diseases of this part of the brain.

GENERAL QUESTIONS

1. What is meant by the terms upper motor neuron and lower motor neuron? Contrast the effects of lower motor neuron lesions with those of lesions affecting each of the types of upper motor neurons.

2. What is encephalization? Discuss its significance.

3. Where are voluntary movements initiated, and how are they brought about?

4. What is the Babinski sign? What is its physiologic and pathologic significance?

5. Describe the mass reflex, a distressing late complication of injuries that transect the spinal cord, and explain why it occurs.

6. What is the physiologic significance of decerebrate rigidity? Where in the nervous system are righting reflexes integrated, and what is their function?

7. What is the evidence that the placing and hopping reflexes are cortical in origin?

8. Define athetosis, ballism, and chorea, and describe the disease processes that produce each of them.

9. List 3 drugs used in the treatment of Parkinson's disease, and explain why each is of value.

10. List 5 types of neurons found in the cerebellar cortex, and describe the morphology and function of each.

11. What is an intention tremor? Why does it occur in cerebellar disease?

MULTIPLE-CHOICE QUESTIONS

In the following questions, select the single best answer.

1. The basal ganglia are primarily concerned with—
    a. sensory integration.
    b. short-term memory.
    c. planning voluntary movement.
    d. neuroendocrine control.
    e. slow-wave sleep.

2. Interruption of the motor pathways in the internal capsule on one side of the body causes—
    a. spastic paralysis on the same side of the body.
    b. spastic paralysis on the opposite side of the body.
    c. loss of touch and pressure sensation on the opposite side of the body.
    d. flaccid paralysis on the same side of the body.
    e. flaccid paralysis on the opposite side of the body.

3. Increased neural activity before a skilled voluntary movement is *first* seen in the—
    a. spinal motor neurons.
    b. precentral motor cortex.
    c. basal ganglia.
    d. cerebellum.
    e. cortical association areas.

4. After falling down a flight of stairs, a young woman is found to have partial loss of voluntary movement on the right side of her body and loss of pain and temperature sensation on the left side below the midthoracic region. It is probable that she has a lesion—
   a. transecting the left half of the spinal cord in the lumbar region.
   b. transecting the left half of the spinal cord in the upper thoracic region.
   c. transecting sensory and motor pathways on the right side of the pons.
   d. transecting the right half of the spinal cord in the upper thoracic region.
   e. transecting the dorsal half of the spinal cord in the upper thoracic region.

In the following questions, one or more than one of the answers may be correct. Select—

A if (1), (2), and (3) are correct;
B if (1) and (3) are correct;
C if (2) and (4) are correct;
D if only (4) is correct; and
E if all are correct.

5. In an athlete catching a ball, neurons in the following regions would be expected to show increased activity—
   (1) the cerebellar cortex.
   (2) the dentate nucleus of the cerebellum.
   (3) area 4 of the cerebral cortex.
   (4) the ventral horns in the spinal cord.

   A   B   C   D   E

6. A positive supporting reaction (magnet reaction)—
   (1) is a reflex mechanism by which spinal animals can be induced to stand.
   (2) is exaggerated in the first 30 minutes after cord transection.
   (3) is due to hyperactive stretch reflexes.
   (4) is inhibited by stimulation of the vestibular nuclei.

   A   B   C   D   E

7. The continuous contraction of antigravity muscles that occurs during standing depends on—
   (1) tonically increased discharge in the γ efferent neurons to the spindles in these muscles.
   (2) increased discharge in the α efferent neurons to the extrafusal fibers in these muscles.
   (3) increased discharge of neurons in the brain stem.
   (4) an intact cerebellum.

   A   B   C   D   E

8. In humans, many of the axons that form the corticospinal tracts—
   (1) end on Renshaw cells.
   (2) end on motor neurons in the spinal cord that supply extrafusal fibers in skeletal muscle.
   (3) end on motor neurons in the spinal cord that innervate muscle spindles.
   (4) end on interneurons in the spinal cord.

   A   B   C   D   E

9. Quadriplegic men frequently have a negative nitrogen balance because—
    (1) they are prone to infections, which in turn increase ACTH secretion.
    (2) they develop hypercalcemia, and this causes dissolution of the protein in bone.
    (3) the muscles of their limbs are paralyzed.
    (4) they lack the afferent input that normally maintains growth hormone secretion.

    A   B   C   D   E

The following 4 questions refer to Fig 12–1. In each, lettered items may be selected once, more than once, or not at all.

10. From which of the labeled structures in Fig 12–1 do the corticospinal and corticobulbar tracts originate?

    A   B   C   D   E

11. Which of the labeled structures in Fig 12–1 is (are) concerned with elaboration of somatic sensory perception?

    A   B   C   D   E

12. Which of the labeled structures in Fig 12–1 is (are) concerned with primary perception of olfactory stimuli?

    A   B   C   D   E

13. Which of the labeled structures in Fig 12–1 is (are) association areas?

    A   B   C   D   E

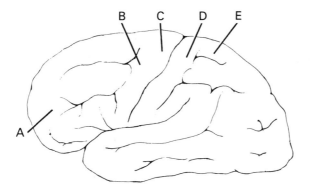

Figure 12–1. Lateral view of human cerebral cortex.

In the following questions, match each numbered item with the lettered items below that are closely associated with it. Lettered items may be selected once, more than once, or not at all.

      (A) Spinal cord.
      (B) Medulla oblongata.
      (C) Hypothalamus.
      (D) Limbic system.

14. Body on head righting reflexes   **A B C D**

15. Tonic neck reflexes   **A B C D**

16. Temperature-regulating reflexes   **A B C D**

17. Withdrawal reflexes   **A B C D**

18. Emotional responses   **A B C D**

In the following questions, indicate whether the item on the left is greater than (**G**), the same as (**S**), or less than (**L**) the item on the right.

19. Frequency of action potentials in afferent fibers from a spindle in a stretched leg muscle of a monkey 1 hour after decortication.   **G S L**   Frequency of action potentials in afferent fibers from a spindle in a stretched leg muscle of a monkey 1 hour after transection of the spinal cord in the upper thoracic region.

20. In a decerebrate animal, frequency of action potentials in motor nerves to limb extensor muscles when the head is turned toward the muscles.   **G S L**   In a decerebrate animal, frequency of action potentials in motor nerves to limb extensor muscles when the head is turned away from the muscles.

21. Motor deficit following decortication in a dog.   **G S L**   Motor deficit following decortication in a monkey.

22. Antiparkinson effect of drug that increases acetylcholine in basal ganglia.   **G S L**   Antiparkinson effect of drug that increases dopamine in basal ganglia.

23. Motor abnormalities at rest in a patient with disease of the cerebellum.   **G S L**   Motor abnormalities at rest in a patient with disease of the basal ganglia.

24. Severity of tremor due to cerebellar disease during voluntary movement.   **G S L**   Severity of tremor due to disease of the basal ganglia during voluntary movement.

REFERENCES

Brooks VB (editor): *Motor Control*. Section 1, Vol 2, of: *Handbook of Physiology*. American Physiological Society, 1981.

Evarts EV: Brain mechanisms of movement. *Sci Am* (Sept) 1979;**241**:146.

The Autonomic Nervous System

13

Chapter 13 is concerned with the motor component of the autonomic nervous system, the efferent pathways from the brain and spinal cord to visceral structures. Afferents in the autonomic nervous system are discussed in Chapters 5 and 7 and central regulation of visceral function in Chapter 14. Chapter 13 describes the preganglionic and postganglionic neurons in the sympathetic and parasympathetic divisions of the autonomic nervous system and discusses the transmitters they secrete as well as the various types of receptors that mediate responses to these transmitters. The responses of autonomic effectors to transmitters are tabulated on the basis of transmitters and receptor types, and the general principles of autonomic physiology and pharmacology are presented.

OBJECTIVES

The material in the chapter should help students to—

- Describe the location of preganglionic sympathetic and parasympathetic neurons in the central nervous system, and identify the nerves by which their axons leave the central nervous system.

- Describe the location of postganglionic sympathetic and parasympathetic neurons and the pathways they take to the visceral structures they innervate.

- Separate the autonomic nervous system into divisions on the basis of the neurotransmitters secreted by the neurons in it, and compare this classification to the anatomic classification into sympathetic and parasympathetic divisions.

- Describe the various potentials produced in postganglionic sympathetic neurons by stimulation of preganglionic neurons, and list the neurotransmitter believed to be responsible for each.

- Outline the functions of the autonomic nervous system.

- List the ways that drugs act to increase or decrease sympathetic activity, parasympathetic activity, or both.

GENERAL QUESTIONS

1. Cannon called mass sympathetic discharge the "preparation for flight or fight." How does sympathetic discharge prepare the individual for flight or fight? Does the sympathetic nervous system have any other functions?

2. Where—outside of the central nervous system—are the cells that secrete catecholamines, and which catecholamine do they secrete?

3. The cholinergic division of the autonomic nervous system has been called the anabolic nervous system. Discuss the actions of the cholinergic division that justify this label.

4. Where are the cell bodies of the postganglionic sympathetic neurons that supply the head? How do they reach the visceral effectors they innervate?

5. Most of the so-called nerve gases are potent inhibitors of acetylcholines-terase. How does inhibition of acetylcholinesterase cause death? What antidote would you give a patient exposed to nerve gas, and why?

6. What neurotransmitters or hormones are secreted by each of the following?
   a. preganglionic autonomic neurons.
   b. postganglionic sympathetic neurons innervating intestinal smooth muscle.
   c. postganglionic sympathetic neurons innervating sweat glands.
   d. postganglionic parasympathetic neurons.
   e. small, intensely fluorescent (SIF) cells.
   f. adrenal medulla.

MULTIPLE-CHOICE QUESTIONS

In the following questions, select the single best answer.

1. Complete denervation of the small intestine would be expected to—
   a. increase the secretion of gastrointestinal hormones.
   b. decrease the resting rate of intestinal peristalsis.
   c. have little effect on the resting rate of intestinal peristalsis.
   d. cause intestinal obstruction.
   e. cause peristalsis to become chaotic and irregular.

2. Which of the following drugs would *not* be expected to increase sympathetic discharge or mimic the effects of increased sympathetic discharge?
   a. norepinephrine.
   b. neostigmine.
   c. amphetamine.
   d. isoproterenol.
   e. atropine.

In the following questions, one or more than one of the answers may be correct. Select—

A if (1), (2), and (3) are correct;
B if (1) and (3) are correct;
C if (2) and (4) are correct;
D if only (4) is correct; and
E if all are correct.

3. Administration of physostigmine, a drug that inhibits acetylcholinesterase, would be expected to—
   (1) increase the secretion of gastric juice.
   (2) increase the rate of melatonin synthesis and secretion from the pineal gland.
   (3) increase the rate of epinephrine secretion from the adrenal medulla.
   (4) decrease the secretion of glucagon.

   A   B   C   D   E

4. In low doses, nicotine—
   (1) acts directly on intestinal smooth muscle, causing it to contract.
   (2) increases the rate of discharge of postganglionic parasympathetic neurons.

(3) increases the release of acetylcholine in sympathetic ganglia.
(4) increases the rate of discharge of postganglionic sympathetic neurons.

A   B   C   D   E

5. In the parasympathetic nervous system—
   (1) the axons of the preganglionic neurons are in the dorsal roots of the spinal nerves in the sacral region.
   (2) the axons of the preganglionic neurons are in the ventral roots of the spinal nerves in the lower thoracic and lumbar regions.
   (3) dopamine is the neurotransmitter responsible for transmission from preganglionic neurons to postganglionic neurons.
   (4) the axons of postganglionic neurons are usually shorter than those of preganglionic neurons.

A   B   C   D   E

6. Administration of a drug that blocks conversion of L-dopa to dopamine would be expected to—
   (1) increase the diameter of the trachea and bronchi.
   (2) disrupt the function of the SIF cells in the sympathetic ganglia.
   (3) decrease peristaltic activity in the small intestine.
   (4) decrease the amount of norepinephrine in the circulating blood.

A   B   C   D   E

7. Administration of a drug that blocks β-adrenergic receptors would be expected to—
   (1) decrease the heart rate.
   (2) decrease the force of cardiac contraction.
   (3) decrease the secretion of renin from the kidneys.
   (4) decrease the secretion of insulin from the B cells in the pancreatic islets.

A   B   C   D   E

8. Polypeptides found in the autonomic nervous system include—
   (1) luteinizing-hormone-releasing hormone (LHRH).
   (2) neuropeptide Y.
   (3) gastrin-releasing peptide (GRP).
   (4) vasoactive intestinal polypeptide (VIP).

A   B   C   D   E

In the following questions, indicate whether the item on the left is greater than **(G)**, the same as **(S)**, or less than **(L)** the item on the right.

9. Number of C fibers in white ramus communicans.   G S L   Number of C fibers in gray ramus communicans.

10. Number of B fibers in white ramus communicans.   G S L   Number of B fibers in gray ramus communicans.

11.	Heart rate when norepinephrine binds to β-adrenergic receptors in the heart.	**G**	**S**	**L**	Heart rate when acetylcholine binds to cholinergic receptors in the heart.
12.	Heart rate when norepinephrine binds to β-adrenergic receptors in the heart.	**G**	**S**	**L**	Heart rate when norepinephrine binds to α-adrenergic receptors in the heart.
13.	Rate of peristalsis when noradrenergic nerves to intestines are stimulated.	**G**	**S**	**L**	Rate of peristalsis when cholinergic nerves to intestines are stimulated.
14.	Rate of insulin secretion when noradrenergic nerves to the pancreas are stimulated.	**G**	**S**	**L**	Rate of insulin secretion when cholinergic nerves to the pancreas are stimulated.
15.	Degree of contraction of bronchial smooth muscle when acetylcholine is injected.	**G**	**S**	**L**	Degree of contraction of bronchial smooth muscle when epinephrine is injected.
16.	Blood glucose concentration in a student taking an examination.	**G**	**S**	**L**	Blood glucose concentration in a student who has been sleeping for 4 hours.

REFERENCES

Weiner N., Taylor P: Neurohumoral transmission: The autonomic and somatic motor nervous system. Chapter 4 in: *Goodman and Gilman's The Pharmacological Basis of Therapeutics,* 7th ed. Goodman AG et al (editors). Macmillan, 1985.

Central Regulation of Visceral Function

14

In the same way that Chapter 12 reviews integration of somatomotor function at various levels in the central nervous system, Chapter 14 reviews the autonomic and related visceral responses integrated in the spinal cord, the medulla oblongata, and the hypothalamus. At the level of the spinal cord, such responses include urination, defecation, and gross control of blood pressure. At the level of the medulla oblongata, they include fine control of blood pressure as well as regulation of heart rate, respiration, swallowing, coughing, and vomiting. At the level of the hypothalamus, they include autonomic responses to emotional stimuli and control of body rhythms, food intake, water intake, body temperature, and endocrine function. In this last context, hypothalamic hormones and control of secretion from the anterior and posterior lobes of the pituitary gland are considered in detail.

OBJECTIVES

The material in the chapter should help students to—

- Describe the autonomic reflexes integrated at the level of the spinal cord.

- Describe vomiting, including its autonomic and somatic components, and discuss the initiation of the vomiting reflex by afferents from the gastrointestinal tract and the chemoreceptor trigger zone.

- Describe the anatomic connections between the hypothalamus and the pituitary gland and the functional significance of each connection.

- List the principal neural pathways from the hypothalamus to other parts of the brain.

- Discuss the relationship of the hypothalamus to sleep and to body rhythms.

- Name the parts of the hypothalamus involved in regulation of food intake, and describe the way they interact to control appetite.

- List the factors that control water intake, and outline the way they exert their effects.

- Discuss the synthesis, processing, storage, and secretion of the hormones of the posterior pituitary.

- List the effects of vasopressin, and discuss how vasopressin secretion is regulated.

- List the effects of oxytocin, and discuss how oxytocin secretion is regulated.

- Name the hypophyseotropic hormones, and outline the effects that each has on anterior pituitary function.

- List the mechanisms by which heat is produced in and lost from the body, and comment on the differences in temperature in the hypothalamus, rectum, oral cavity, and skin.

- List the temperature-regulating mechanisms, and describe the way they are integrated under hypothalamic control to maintain normal body temperature.

- Discuss the pathophysiology of fever.

GENERAL QUESTIONS

1. What is interleukin-1 (endogenous pyrogen)? Describe its production and discuss its effects.

2. What is behavioral thermoregulation? Give examples and describe how they operate to maintain body temperature.

3. What functions do the hypophyseotropic hormones have in addition to regulation of anterior pituitary secretion?

4. What endocrine changes would you expect to see after section of the pituitary stalk, and why?

5. Draw graphs showing the relation of thirst to plasma osmolality, plasma vasopressin to plasma osmolality, and plasma vasopressin to extracellular fluid volume.

6. How do changes in extracellular fluid volume affect vasopressin secretion?

7. How do the various factors that affect vasopressin secretion interact, eg, how does a decline in extracellular fluid volume affect the threshold for osmotic stimulation of vasopressin secretion?

8. Why does food intake increase in untreated diabetes mellitus?

9. Using the axes in Fig 14–1, draw the curve relating metabolic rate to body temperature from 95 °F to 106 °F (35–41 °C).

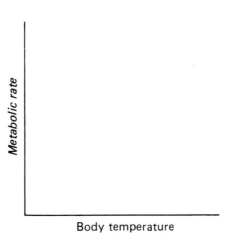

Body temperature

Figure 14–1.

10. Some individuals drink large quantities of water for psychologic reasons (psychogenic polydipsia). How would you differentiate between psychogenic polydipsia, diabetes insipidus due to a hypothalamic lesion, and diabetes insipidus due to an inability of the kidneys to respond to vasopressin (nephrogenic diabetes insipidus)?

11. What is the effect of bilateral lesions of the area postrema on the vomiting reflex?

12. Discuss the long-term regulation of appetite in terms of catch-up growth and the response to force-feeding.

MULTIPLE-CHOICE QUESTIONS

In the following questions, select the single best answer.

1. Thirst is stimulated by—
   a. increases in plasma osmolality and volume.
   b. an increase in plasma osmolality and a decrease in plasma volume.
   c. a decrease in plasma osmolality and an increase in plasma volume.
   d. decreases in plasma osmolality and volume.
   e. injection of vasopressin into the hypothalamus.

2. The appetite for food is *not* increased by—
   a. hypoglycemia.
   b. decreased glucose utilization in the ventromedial nuclei.
   c. insulin deficiency.
   d. distention of the stomach.
   e. a fall in body temperature.

3. When an individual is standing naked in a room in which the air temperature is 21 °C (69.8 °F) and the humidity 80%, the greatest amount of heat will be lost from the body by—
   a. urination.
   b. respiration.
   c. elevated metabolism.
   d. vaporization of sweat.
   e. radiation and conduction.

The following 5 questions refer to Fig 14–2. In each, lettered items may be selected once, more than once, or not at all.

4. In which of the labeled structures in Fig 14–2 is there a high concentration of vasopressin?

A   B   C   D   E

5. In which of the labeled structures in Fig 14–2 is there a high concentration of corticotropin-releasing hormone (CRH)?

A   B   C   D   E

6. In which of the labeled structures in Fig 14–2 is there a high concentration of thyrotropin-releasing hormone (TRH)?

A   B   C   D   E

7. In which of the labeled structures in Fig 14–2 is there a high concentration of dopamine?

A   B   C   D   E

8. In which of the labeled structures in Fig 14–2 is there a high concentration of glucoreceptors?

A   B   C   D   E

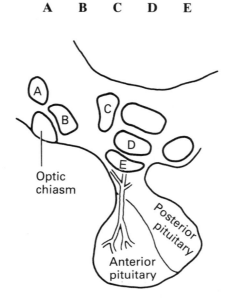

Figure 14–2. Hypothalamus and pituitary.

In the following questions, select—
> **A** if the item is associated with **(a)** below,
> **B** if the item is associated with **(b)** below,
> **C** if the item is associated with both **(a)** and **(b)**, and
> **D** if the item is associated with neither **(a)** nor **(b)**.

   (a) V_1 vasopressin receptors.
   (b) V_2 vasopressin receptors.

9. Activation of G_s **A   B   C   D**

10. Vasoconstriction.   **A   B   C   D**

11. Increase in intracellular inositol triphosphate.   **A   B   C   D**

12. Antidiuresis.   **A   B   C   D**

13. Glycosuria.   **A   B   C   D**

In the following questions, one or more than one of the answers may be correct. Select—
> **A** if (1), (2), and (3) are correct;
> **B** if (1) and (3) are correct;
> **C** if (2) and (4) are correct;
> **D** if only (4) is correct; and
> **E** if all are correct.

14. Day-to-day regulation of water intake is dependent on—
   (1) stretch receptors in the atria.
   (2) chemoreceptors in the supraoptic nuclei.
   (3) osmoreceptors in the circumventricular organs.
   (4) chemoreceptors in the carotid bodies.

   A   B   C   D   E

15. If one leg is immersed in ice water, the subject's immediate responses include—
   (1) generalized vasoconstriction.
   (2) increased secretion of epinephrine.
   (3) increased blood pressure.
   (4) shivering.

   A   B   C   D   E

16. Diabetes insipidus is characterized by—
   (1) excretion of large volumes of concentrated urine.
   (2) a paradoxic increase in the plasma vasopressin concentration.
   (3) chronically depressed plasma osmolality.
   (4) excretion of large volumes of dilute urine.

   A   B   C   D   E

17. Which of the following hypothalamic hypophyseotropic hormones have an established effect on the secretion of more than one anterior pituitary hormone?
   (1) TRH.
   (2) LHRH.
   (3) CRH.
   (4) growth-hormone releasing hormone (GRH).

   A   B   C   D   E

18. Circulating angiotensin II stimulates adrenocorticotropic hormone (ACTH) secretion but does not cross the blood-brain barrier and apparently does not act directly on the anterior pituitary gland in vivo. Therefore, it is reasonable to hypothesize that it acts on the—
   (1) organum vasculosum of the lamina terminalis.
   (2) paraventricular nuclei.
   (3) subfornical organ.
   (4) arcuate nuclei.

   A   B   C   D   E

19. Vomiting can be initiated by—
   (1) injection of apomorphine and related emetic drugs.
   (2) irritation of the gastric mucosa.

(3) the sight and smell of rotten food.

(4) injection of drugs that block dopamine receptors.

A B C D E

20. The following lower body temperature—

  (1) curling up.

  (2) decreased food intake.

  (3) increased circulating epinephrine.

  (4) increased respiration.

A B C D E

In the following questions, indicate whether the item on the left is greater than **(G)**, the same as **(S)**, or less than **(L)** the item on the right.

21. Body temperature in hyper-thyroidism. **G** **S** **L** Body temperature in hypo-thyroidism.

22. In a normal subject on a usual schedule of daytime and nighttime activities, body temperature in the morning. **G** **S** **L** In the same subject, body temperature in the afternoon.

23. Osmolality of present-day seawater. **G** **S** **L** Osmolality of extracellular fluid.

24. Appetite after lateral hypo-thalamic lesions. **G** **S** **L** Appetite after medial hypo-thalamic lesions.

25. Plasma oxytocin concentration before suckling an infant. **G** **S** **L** Plasma oxytocin concentration immediately after suckling an infant.

REFERENCES

Bennett GW, Whitehead SA: *Mammalian Neuroendocrinology*. Oxford Univ Press, 1983.

Reichlin S, Baldessarini RJ, Martin JB (editors): *The Hypothalamus*. Raven Press, 1978.

Neural Basis of Instinctual Behavior & Emotions

15

Chapter 15 is concerned with the limbic system and with the neural circuits involved in emotional responses and stereotyped behaviors. These include sexual and maternal behavior, fear, rage, and motivation. The neurons in the brain that secrete dopamine, norepinephrine, serotonin, acetylcholine, opioids, and gamma-aminobutyric acid (GABA) are mapped and their functions reviewed with special emphasis on their relation to emotional behavior.

OBJECTIVES

The material in the chapter should help students to—

- Describe in general terms the structure and function of the limbic system.

- Discuss the brain regions and hormones involved in the regulation of sexual behavior in both sexes.

- Summarize the effects of sex hormones on the brain in fetal and early neonatal life.

- Describe the parts of the brain involved in producing the balance between rage and placidity.

- Describe the brain systems that mediate repeated self-stimulation and avoidance of stimulation.

- Outline the anatomy of the serotonergic pathways, and summarize their known and suspected functions.

- Outline the anatomy of the noradrenergic (norepinephrine-secreting) pathways in the brain, and summarize their known and suspected functions.

- Outline the anatomy of the adrenergic (epinephrine-secreting) pathways in the brain, and summarize their known and suspected functions.

- Outline the anatomy of the dopaminergic pathways in the brain, and summarize their known and suspected functions.

- Outline the opioid peptide-secreting pathways in the brain, and summarize their known and suspected functions.

GENERAL QUESTIONS

1. Discuss sham rage. How is it produced? What is the evidence indicating that the word sham is inappropriate?

2. Compare the hyperphagia produced by lesions of the amygdaloid nuclei to that produced by lesions of the ventromedial nuclei of the hypothalamus.

3. What is a pheromone? What are the sources and functions of pheromones in monkeys and humans?

4. Discuss the role of ovarian and adenocortical hormones in the regulation of sexual behavior in women.

5. Some scientists talk about the "female hypothalamus" and the "male hypothalamus." What are the differences between the hypothalami of the 2 sexes, and how are these differences produced?

6. What emotional changes are produced in experimental animals and humans by bilateral lesions of the amygdaloid nuclei?

7. Discuss the relationship between brain amine transmitters and sleep.

8. What is the putative relationship between norepinephrine and mood? What pharmacologic agents have been used to treat depression, and why?

9. What are the relationships of brain dopamine and brain serotonin to mental abnormalities and other diseases?

MULTIPLE-CHOICE QUESTIONS

In the following questions, select the single best answer. Each question is made up of a statement and an explanation. Select—
- **A** if both the statement and the explanation are true and they are related;
- **B** if both the statement and the explanation are true but they are unrelated;
- **C** if the statement is true but the explanation is false;
- **D** if the statement is false but the explanation is true; and
- **E** if the statement and the explanation are both false.

1. L-Dopa is of benefit in the treatment of Parkinson's disease because it increases the dopamine content of the tuberoinfundibular dopaminergic neurons.

    A   B   C   D   E

2. Drugs that decrease norepinephrine reuptake are probably of benefit in the treatment of some depressive illnesses because they increase extracellular norepinephrine in the brain.

    A   B   C   D   E

3. Procedures that increase the release of opioid peptides in the central nervous system decrease the effect of painful stimuli because the periaqueductal gray and related areas contain opioid receptors to which the peptides can bind.

    A   B   C   D   E

4. The antipsychotic activity of the major tranquilizers parallels their ability to block serotonin receptors because the serotonergic neural systems in the brain are concerned among other things with normal mental function.

    A   B   C   D   E

5. Women with a congenital adrenocortical enzyme defect that causes large amounts of androgens to be secreted have somewhat more masculine behavior than normal women because the androgens act on the brain during fetal development.

A   B   C   D   E

6. "Sham rage" is sometimes seen in patients after encephalitis or brain surgery because these conditions can damage pathways from the neocortex to the limbic system.

A   B   C   D   E

In the following questions, one or more than one of the answers may be correct. Select—

A if (1), (2), and (3) are correct;
B if (1) and (3) are correct;
C if (2) and (4) are correct;
D if only (4) is correct; and
E if all are correct.

7. Which of the following are part of the limbic system?
   (1) the red nuclei.
   (2) the amygdaloid nuclei.
   (3) the posterior commissure.
   (4) the cingulate cortex.

A   B   C   D   E

8. The functions of the limbic system include—
   (1) regulation of sexual behavior in the male.
   (2) expression of fear.
   (3) olfaction.
   (4) comprehension of language.

A   B   C   D   E

9. Rage behavior can be—
   (1) produced by lesions of the septal nuclei.
   (2) increased by administration of male hormones.
   (3) produced by stimulation of the hypothalamus.
   (4) produced by lesions of the hypothalamus.

A   B   C   D   E

10. Cholinergic neurons are found in the brain in—
    (1) the basal ganglia.
    (2) the neocortex.
    (3) the nucleus basalis of Meynert.
    (4) the spinal cord.

A   B   C   D   E

11. The following would be expected to reduce maternal behavior—
    (1) lesions of the cingulate cortex.
    (2) lesions of the mediobasal hypothalamus.
    (3) hypophysectomy.
    (4) ovariectomy.

A   B   C   D   E

12. Benzodiazepines such as diazepam (Valium) relieve anxiety because they—
    (1) inhibit dopamine binding in the brain stem.
    (2) inhibit glutamate binding in the limbic system.
    (3) facilitate norepinephrine release in the hypothalamus.
    (4) facilitate the effects of GABA throughout the nervous system.

A   B   C   D   E

13. Which of the following hallucinogenic drugs appear to produce their effects via serotonin receptors?
    (1) lysergic acid diethylamide (LSD).
    (2) psilocin.
    (3) N,N-dimethyltryptamine (DMT).
    (4) mescaline.

A   B   C   D   E

In the following questions, indicate whether the item on the left is greater than **(G)**, the same as **(S)**, or less than **(L)** the item on the right.

14. Number of points in the brain where animals will avoid stimulation through an implanted electrode.   **G S L**   Number of points in the brain where animals will work to receive repeated stimulation through an implanted electrode.

15. Number of points in the brain where animals will work to receive repeated stimulation through an implanted electrode.   **G S L**   Number of points in the brain where animals' response to stimulation through an implanted electrode is indifferent, ie, the animals neither seek nor avoid stimulation.

16. Amount of norepinephrine in the locus ceruleus.   **G S L**   Amount of dopamine in the locus ceruleus.

17. Amount of norepinephrine in the raphe nuclei.   **G S L**   Amount of serotonin in the raphe nuclei.

18. Amount of norepinephrine in the nigrostriatal pathway.   **G S L**   Amount of dopamine in the nigrostriatal pathway.

19. Amount of dopamine in tyrosine hydroxylase-containing neurons in the arcuate nuclei of the hypothalamus.

G S L

Amount of dopamine in tyrosine hydroxylase-containing neurons in the pre-optic region of the hypothalamus.

REFERENCES

Isaacson RL: *The Limbic System.* Plenum, 1974.
Kandel ER, Schwartz JH (editors): *Principles of Neural Science,* 2nd ed. Chapters 37 and 38. Elsevier, 1985.

16

"Higher Functions of the Nervous System": Conditioned Reflexes, Learning, & Related Phenomena

The physiology of learning and memory is considered in Chapter 16, and habituation, sensitization, conditioned reflexes, and the various types of memory are discussed. Complementary specialization of the cerebral hemispheres is described, along with its relation to handedness. The language functions of the neocortex are also analyzed, and other functions of the frontal and temporal lobes are reviewed.

OBJECTIVES

The material in the chapter should help students to—

- Give examples of habituation and sensitization occurring in humans, and discuss the molecular events that may underlie them.

- Describe the evoked potentials produced in the brain by a neural stimulus the first time it is applied, after it has been applied repeatedly, and after it has been paired a number of times with an unpleasant electric shock.

- Define conditioned reflexes, describe their properties, and analyze their physiologic basis.

- List and explain the abnormalities of brain function produced by sectioning the corpus callosum.

- List the parts of the brain that appear to be involved in memory in mammals, and summarize the proposed role of each in memory processing and storage.

- Describe the abnormalities of brain structure and function found in Alzheimer's disease.

- Describe the functions of the association areas in the cerebral cortex.

- Define the terms categorical hemisphere and representational hemisphere, and summarize the differences between the hemispheres and their relationship to handedness.

- Define and explain agnosia, unilateral neglect, dyslexia, and prosopagnosia.

- Summarize the differences between fluent and nonfluent aphasias, and explain each type on the basis of its pathophysiology.

- Describe the effects of frontal lobectomy and temporal lobectomy on behavior.

GENERAL QUESTIONS

1. What methods are used to study learning and memory?

2. Give examples of learning that occurs in invertebrates with simple nervous systems. What is the neural and biochemical basis of this learning?

3. What is the orienting reflex?

4. Why does a mother often sleep through many different kinds of noise but wake when her baby cries?

5. What is biofeedback? Explain it in physiologic terms.

6. What is discriminative conditioning? What are delayed conditioned reflexes and conditioned avoidance reflexes?

7. List and explain the differences between declarative and reflexive memory.

8. It is now established that the 2 cerebral hemispheres are not anatomically symmetrical. What are the anatomic differences between the 2 sides, and what is their relation to the physiologic differences between the hemispheres?

9. Discuss the functions of Wernicke's area, Broca's area, and the angular gyrus in the production of speech.

10. What are the effects of prefrontal lobotomy? Contrast them with the effects of bilateral ablation of the temporal lobes.

MULTIPLE-CHOICE QUESTIONS

In the following questions, select the single best answer.

1. Retrograde amnesia—
   a. is abolished by prefrontal lobotomy.
   b. responds to drugs that block dopamine receptors.
   c. is commonly precipitated by a blow on the head.
   d. is increased by administration of vasopressin.
   e. is due to damage to the brain stem.

2. The representational hemisphere is—
   a. the right cerebral hemisphere in most right-handed individuals.
   b. the left cerebral hemisphere in most left-handed individuals.
   c. the part of the brain concerned with language functions.
   d. the site of lesions in most patients with aphasia.
   e. morphologically identical to the opposite nonrepresentational hemisphere.

3. The optic chiasm is sectioned in a dog, and with the right eye closed, the animal is trained to bark when it sees a red square. The right eye is then opened and the left eye closed. The animal will now—
   a. fail to respond to the red square because it does not produce impulses that reach the right occipital cortex.
   b. fail to respond to the red square because the animal has bitemporal hemianopia.
   c. fail to respond to the red square if the posterior commissure is also sectioned.

    d. respond to the red square only after retraining.
    e. respond promptly to the red square in spite of the lack of input to the left occipital cortex.

4. The effects of bilateral loss of hippocampal function include—
    a. disappearance of remote memories.
    b. inability to recall the events of the moment.
    c. loss of short-term memory, ie, inability to encode events of the recent past in long-term memory.
    d. loss of the ability to recall faces and forms but not the ability to recall printed or spoken words.
    e. production of inappropriate emotional responses when recalling events of the recent past.

5. Which of the following are *incorrectly* paired?
    a. lesion of the inferior parietal lobule of the representational hemisphere : unilateral inattention and neglect.
    b. loss of cholinergic neurons in the nucleus basalis of Meynert and related areas of the forebrain : loss of recent memory.
    c. lesions of mamillary bodies : loss of recent memory.
    d. lesion of the angular gyrus in the categorical hemisphere : nonfluent aphasia.
    e. lesion of Broca's area in the categorical hemisphere : automatic words.

In the following questions, one or more than one of the answers may be correct. Select—

        A if (1), (2), and (3) are correct;
        B if (1) and (3) are correct;
        C if (2) and (4) are correct;
        D if only (4) is correct; and
        E if all are correct.

6. Bilateral temporal lobectomy in monkeys causes—
    (1) hyperphagia.
    (2) decreased sexual activity in females.
    (3) increased sexual activity in males.
    (4) weight loss.

        A   **B**   **C**   **D**   **E**

7. The representational hemisphere is better than the categorical hemisphere at—
    (1) language functions.
    (2) recognition of objects by their form.
    (3) understanding printed words.
    (4) identification of musical themes.

        A   **B**   **C**   **D**   **E**

8. Which of the following are characteristic of conditioned reflexes?
    (1) failure to form conditioned reflexes when the conditioned and unconditioned stimulus are separated by more than 2 minutes.
    (2) disappearance of the conditioned reflex if it is not reinforced from time to time.
    (3) rapid formation of conditioned reflexes when the conditioned stimulus is pleasant for the animal.

(4) failure to form conditioned reflexes when the conditioned stimulus is unpleasant to the animal.

A　　B　　C　　D　　E

9. Discriminative conditioning can be used to—
   (1) obliterate undesirable conditioned reflexes.
   (2) determine thresholds for sensory discrimination.
   (3) study the causes of aphasia.
   (4) produce the experimental neurosis.

A　　B　　C　　D　　E

10. A lesion of Wernicke's area (the posterior end of the superior temporal gyrus) in the categorical hemisphere causes patients to—
    (1) lose short-term memory.
    (2) speak in a slow, halting voice.
    (3) experience the déjà vu phenomenon.
    (4) talk rapidly but make little sense.

A　　B　　C　　D　　E

In the following questions, match each numbered item with the lettered items below that are closely associated with it. Lettered items may be selected once, more than once, or not at all.
   (A) Experimental neurosis.
   (B) Short-term memory.
   (C) Motor function.
   (D) Inability to recognize faces.

11. Hypothalamus　　　　　　　　　　　A　B　C　D

12. Frontal lobe　　　　　　　　　　　　A　B　C　D

13. Hippocampus　　　　　　　　　　　A　B　C　D

14. Temporal lobe　　　　　　　　　　　A　B　C　D

15. Basal ganglia　　　　　　　　　　　A　B　C　D

16. Amygdala　　　　　　　　　　　　　A　B　C　D

In the following questions, indicate whether the item on the left is greater than **(G)**, the same as **(S)**, or less than **(L)** the item on the right.

17. Size of the superior tempo-　　G　S　L　　Size of the superior tempo-
    ral gyrus in the representa-　　　　　　　　ral gyrus in the categorical
    tional hemisphere.　　　　　　　　　　　hemisphere.

18. Speech abnormalities fol-　　G　S　L　　Speech abnormalities fol-
    lowing a lesion in the repre-　　　　　　　lowing a lesion in the cate-
    sentational hemisphere.　　　　　　　　gorical hemisphere.

19.	Number of left-handed people in the general population.	G S L	Number of right-handed people in the general population.
20.	Size of evoked potentials in the cortex the first time a subject is exposed to a neutral stimulus.	G S L	Size of evoked potentials in the cortex the 20th time a subject is exposed to a neutral stimulus.
21.	Size of the brain in an adult porpoise.	G S L	Size of the brain in an adult human.
22.	Brain weight per unit of body weight in an adult elephant.	G S L	Brain weight per unit of body weight in an adult human.
23.	Incidence of dyslexia in right-handed individuals.	G S L	Incidence of dyslexia in left-handed individuals.
24.	Emotional reaction to pain before prefrontal lobotomy.	G S L	Emotional reaction to pain after prefrontal lobotomy.

REFERENCES

Geschwind N: Specializations of the human brain. *Sci Am* (Sept) 1979;**241**:158.

Trevarthen C: Hemispheric specialization. In: *The Nervous System*. Section 2, Vol 3 of: *Handbook of Physiology*. American Physiological Society, 1984.

Energy Balance, Metabolism, & Nutrition

17

Chapter 17 provides a summary and overview of energy balance and metabolism as a background for consideration of the functions of the endocrine glands concerned with metabolism. Energy production and utilization are considered, along with the metabolic rate and the factors that affect it. The biochemical reactions involved in energy transformation and storage are summarized, and the major pathways involved in the metabolism of carbohydrate, protein, and fat are reviewed. The topics considered in the analysis of fat metabolism include lipid transport and the production of prostaglandins, prostacyclin, thromboxanes, and leukotrienes from arachidonic acid. The essentials of nutrition, the minimum daily requirements of nutrients, and the physiologic role of vitamins are also reviewed.

OBJECTIVES

The material in the chapter should help students to—

- Define metabolic rate, calorie, and respiratory quotient.

- State the caloric value per unit weight of carbohydrate, protein, and fat.

- Describe how the basal metabolic rate is measured, and list some of the major diseases in which it is abnormal.

- Discuss the biosynthesis and functions of adenosine triphosphate.

- List the principal dietary hexoses, and describe the main pathways for intracellular metabolism of glucose.

- Describe the function of the citric acid cycle in the metabolism of glucose, amino acids, and fatty acids.

- Summarize the main factors and mechanisms that regulate glucose synthesis and breakdown.

- Define the glucostatic function of the liver, and explain how the liver carries out this function.

- Define amino acid, polypeptide, and protein.

- Define gluconeogenesis, urea cycle, creatine, and creatinine.

- Describe the structure of DNA and RNA and the way purines and pyrimidines are joined together to form these molecules.

- List the main sources of uric acid in the body and the main metabolic products produced from uric acid.

- Describe the metabolic responses to starvation.

- List the major classes of lipids in the body, and describe the principal characteristics of each.

- Define ketone bodies, and describe their formation and metabolism.

- Describe the sources of the free fatty acids in plasma, their metabolic fate, and the principal factors regulating the fatty acid level in plasma.

- Describe the exogenous and endogenous pathways by which lipids are transported in the body, and summarize the processes involved in cholesterol metabolism.

- Outline the major pathways involved in the formation of leukotrienes, thromboxanes, prostacyclin, and prostaglandins.

- List the components in a normal diet that will maintain weight and health.

- Define the term vitamin, name the major vitamins, and summarize the effects of deficiency and excess of each vitamin.

GENERAL QUESTIONS

1. The oxygen consumption of a fasting animal at rest was found to be 30 mL/ kg body weight/min. What is the basal metabolic rate (BMR) (in kcal/kg/ 24 h)? What is the approximate size of this animal?

2. What is specific dynamic action? How is it produced?

3. What is a "directional flow valve" in metabolism? Why are such valves important? Name the principal ones.

4. What is alimentary glycosuria? Explain its occurrence.

5. Discuss the changes in carbohydrate metabolism that occur during exercise.

6. Compare the metabolism of fructose with that of glucose. How do they interact?

7. What factors affect the blood level of amino acids?

8. Discuss gout.

9. What are essential amino acids?

10. What treatment would you prescribe to lower plasma cholesterol? Why is cholesterol reduction important? Discuss the relation of cholesterol to low-density lipoproteins (LDL) and high-density lipoproteins (HDL).

11. How are the metabolites of arachidonic acid important in the body? Why has a major effort been devoted to the development of drugs that modify arachidonic acid metabolism?

12. Discuss the pathophysiology of obesity. How would you treat this condition?

MULTIPLE-CHOICE QUESTIONS

In the following questions, select the single best answer.

1. In the body, metabolism of 10 g of protein would produce approximately—

   a. 1 kcal.
   b. 41 kcal.
   c. 410 kcal.
   d. 4100 kcal.
   e. 41 cal.

2. A man with a respiratory quotient (RQ) of 70—
   a. has been eating a high-fat diet.
   b. has been eating a high-protein diet.
   c. has been starving for 48 hours.
   d. has eaten nothing but carbohydrates for 24 hours.
   e. is dehydrated.

3. In the body, metabolism of 10 g of carbohydrate would produce approximately—
   a. 93 kcal.
   b. 9.3 kcal.
   c. 4.1 kcal.
   d. 41 kcal.
   e. 45 kcal.

4. The lipoprotein that is the major source of the cholesterol used in cells is—
   a. chylomicrons.
   b. intermediate-density lipoproteins (IDL).
   c. very-low-density lipoproteins (VLDL).
   d. LDL.
   e. HDL.

5. Which of the following produces the most high-energy phosphate compounds?
   a. aerobic metabolism of 1 mol of glucose.
   b. anaerobic metabolism of 1 mol of glucose.
   c. metabolism of 1 mol of galactose.
   d. metabolism of 1 mol of amino acid.
   e. metabolism of 1 mol of long-chain fatty acid.

6. Which of the following produces the most glucose when metabolized in the body?
   a. 1 mol of serotonin.
   b. 1 mol of alanine.
   c. 1 mol of oleic acid.
   d. 1 mol of thyroxine (T_4).
   e. 1 mol of acetoacetic acid.

7. Which of the following would *not* produce an increase in the plasma level of free fatty acids?
   a. a drug that increases the level of intracellular cyclic AMP.
   b. a drug that activates β_1-adrenergic receptors.
   c. a drug that inhibits hormone-sensitive lipase.
   d. a drug that decreases the metabolic clearance of glucagon.
   e. a drug that inhibits phosphodiesterase.

In the following questions, one or more than one of the answers may be correct. Select—

        A if (1), (2), and (3) are correct;
        B if (1) and (3) are correct;
        C if (2) and (4) are correct;
        D if only (4) is correct; and
        E if all are correct.

8. The metabolic rate is increased after—
   (1) a decrease in the environmental temperature from 27 °C to 7 °C.
   (2) injection of epinephrine.
   (3) consumption of a meal rich in fat.
   (4) consumption of a meal rich in protein.

   A   B   C   D   E

9. The metabolic rate is affected by—
   (1) body size.
   (2) emotional state.
   (3) age.
   (4) the plasma level of triiodothyronine (T_3).

   A   B   C   D   E

10. Gluconeogenesis in the liver is enhanced by increased plasma levels of—
    (1) alanine.
    (2) glucagon.
    (3) cortisol.
    (4) free fatty acids.

    A   B   C   D   E

11. Lipoprotein lipase lowers the plasma concentration of—
    (1) chylomicron remnants.
    (2) chylomicrons.
    (3) HDL.
    (4) VLDL.

    A   B   C   D   E

12. Which of the following are ketoacids?
    (1) acetone.
    (2) β-hydroxybutyrate.
    (3) glycerol.
    (4) acetoacetate.

    A   B   C   D   E

13. Which of the following have a protein-sparing effect?
    (1) fat.
    (2) uric acid.
    (3) glucose.
    (4) T_4.

    A   B   C   D   E

14. When LDL enters cells by receptor-mediated endocytosis, there is—
    (1) an increase in the formation of cholesterol from mevalonic acid.
    (2) a decrease in the intracellular concentration of free cholesterol.
    (3) a decrease in the transfer of cholesterol from the cell to HDL.
    (4) a decrease in the rate of synthesis of LDL receptors.

    A   B   C   D   E

15. The output of glucose from the liver is increased by—
    (1) induction of phosphorylase.
    (2) insulin.
    (3) glucagon.
    (4) fructose 2,6-diphosphate.

A   B   C   D   E

In the following questions, indicate whether the item on the left is greater than **(G)**, the same as **(S)**, or less than **(L)** the item on the right.

16. Caloric value of 1 g of fat.   **G S L**   Caloric value of 1 g of carbohydrate.

17. Absorption of fat-soluble vitamins in a patient with steatorrhea.   **G S L**   Absorption of water-soluble vitamins in a patient with steatorrhea.

18. Amount of ATP generated when glucose is metabolized aerobically.   **G S L**   Amount of ATP generated when the same amount of glucose is metabolized anaerobically.

19. Life expectancy in an individual with low plasma LDL levels.   **G S L**   Life expectancy in an individual with high plasma LDL levels.

20. Loss of nitrogen in the urine in untreated diabetes mellitus.   **G S L**   Loss of nitrogen in the urine in untreated diabetes insipidus.

21. Caloric value of 1 g of protein oxidized outside the body.   **G S L**   Caloric value of 1 g of protein metabolized in the body.

22. Synthesis of leukotrienes after treatment with aspirin.   **G S L**   Synthesis of leukotrienes after treatment with dexamethasone.

23. Minimum daily requirement of iron in an average adult man.   **G S L**   Minimum daily requirement of iron in an average adult woman.

24. Minimum daily requirement of calories in an average adult man.   **G S L**   Minimum daily requirement of calories in an average adult woman.

25. Ability to see at night in vitamin A deficiency.   **G S L**   Ability to see at night in vitamin D deficiency.

26. Hematocrit in a patient with vitamin B_{12} deficiency.   **G S L**   Hematocrit in a patient with vitamin B_1 deficiency.

27. Plasma concentration of free fatty acids 10 minutes after injection of epinephrine.   **G S L**   Plasma concentration of free fatty acids 10 minutes after injection of growth hormone.

28. Amount of NH₃ formed from glutamine in the brain. **G S L** Amount of NH₃ formed from glutamine in the kidneys.

29. Amount of fatty acid synthesis in the mitochondria. **G S L** Amount of fatty acid oxidation in the mitochondria.

30. Plasma uric acid level following treatment of gout with colchicine. **G S L** Plasma uric acid level following treatment of gout with allopurinol.

REFERENCES

Murray RK et al: *Harper's Biochemistry*, 21st ed. Appleton & Lange, 1987.

The Thyroid Gland

18

Chapter 18 reviews the functional anatomy of the thyroid gland and considers in detail the biosynthesis, storage, secretion, binding, metabolism, and effects of T_4 and T_3—the principal thyroid hormones—and the inactive compound reverse triiodothyronine (RT_3). Iodine metabolism is also considered in relation to thyroid function. The regulation of thyroid function is reviewed, and in addition, there is a brief review of the major diseases of the thyroid gland and drugs that affect thyroid function.

OBJECTIVES

The material in the chapter should help students to—

- Describe the gross and microscopic anatomy of the thyroid gland and the effect of thyroid-stimulating hormone (TSH) excess and deficiency on its morphology.

- Compare the structure of T_4, T_3, and RT_3.

- Outline the steps involved in the biosynthesis of thyroid hormones and their storage in the colloid.

- Describe the steps involved in the transfer of thyroid hormones from the colloid to the bloodstream.

- Name the proteins that bind thyroid hormones in the plasma, describe the relationship between bound and free thyroid hormones, and summarize the mechanisms that regulate thyroid hormone binding.

- Outline the principal pathways by which thyroid hormones are metabolized.

- List the main physiologic actions of thyroid hormones.

- Describe the mechanisms by which thyroid hormones exert their principal effects.

- Outline the processes involved in the regulation of secretion of thyroid hormones.

- List the principal drugs that affect thyroid function, and describe the mechanism by which each exerts its effect.

- List the major diseases of the thyroid and their principal symptoms and signs.

GENERAL QUESTIONS

1. The curves in Fig 18–1 show the changes in basal metabolic rate that follow administration of a single dose of T_3 or T_4 to an individual with previously untreated hypothyroidism. Which curve is produced by T_3 and which by T_4? Explain the differences between the 2 curves.

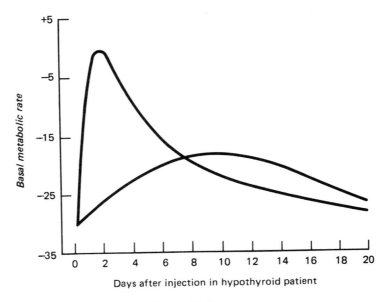

Figure 18–1.

2. What is the role of TRH in the day-to-day regulation of thyroid function?

3. Discuss the causes and the pathophysiology of goiter. Why do people who eat large amounts of cabbage sometimes develop goiters?

4. Why do children with cretinism have short stature and mental deficiency?

5. Patients with thyrotoxicosis often have dramatic remission of many of their signs and symptoms upon administration of β-adrenergic blocking drugs such as propranolol. Which signs and symptoms are improved and which are not? Why?

6. How would you demonstrate that the primary site of feedback of thyroid hormones on TSH secretion is the anterior pituitary rather than the hypothalamus?

7. What hormones produce an increase in metabolic rate? How are these effects produced?

8. How does the thyroid gland take up iodide from the extracellular fluid? What physiologic factors and drugs affect this uptake?

9. Discuss the effects on iodine metabolism of (1) an increase in the glomerular filtration rate, (2) a decrease in uptake of iodine by the thyroid, and (3) a decrease in the amount of iodine in the diet.

10. What are the embryologic origins of the thyroid gland, the parathyroid glands, and the parafollicular cells of the thyroid gland? Why is it important for clinicians to know about the embryologic development of these tissues?

MULTIPLE-CHOICE QUESTIONS

In the following questions, select the single best answer.

1. Propylthiouracil causes—
    a. a decrease in thyroid size and an increase in plasma T_4.
    b. a decrease in both thyroid size and plasma T_4.
    c. an increase in both thyroid size and plasma T_4.
    d. an increase in thyroid size and a decrease in plasma T_4.
    e. no change in thyroid size and a decrease in plasma T_4.

2. A young woman has puffy skin and a hoarse voice. Her plasma TSH concentration is low but increases markedly when she is given TRH. She probably has—
    a. hyperthyroidism due to a thyroid tumor.
    b. hypothyroidism due to a primary abnormality in the thyroid gland.
    c. hypothyroidism due to a primary abnormality in the pituitary gland.
    d. hypothyroidism due to a primary abnormality in the hypothalamus.
    e. hyperthyroidism due to a primary abnormality in the hypothalamus.

3. The coupling of monoiodotyrosine and diiodotyrosine and the iodination of thyroglobulin are blocked by—
    a. divalent cations.
    b. monovalent anions such as perchlorate.
    c. TSH.
    d. TRH.
    e. thiocarbamides such as propylthiouracil.

4. The thyroidal uptake of radioactive iodine is *not* affected by—
    a. calcitonin.
    b. T_3.
    c. TSH.
    d. propylthiouracil.
    e. the iodine content of the diet.

5. The BMR is *least* affected by an increase in the plasma level of—
    a. TSH.
    b. TRH.
    c. thyroxine-binding globulin (TBG).
    d. free T_4.
    e. free T_3.

6. A patient has low thyroidal uptake of radioactive iodine, a high plasma T_4 level, and a normal or low plasma T_3 level. The most likely cause is—
    a. treatment with birth control pills.
    b. a thyroid tumor.
    c. failure of the thyroid gland to form T_3.
    d. treatment with T_4.
    e. a defect in T_3 formation in skeletal muscle and liver.

7. Which of the following is *not* essential for normal biosynthesis of thyroid hormones?
    a. iodine.
    b. ferritin.
    c. thyroglobulin.
    d. protein synthesis.
    e. TSH.

In the following questions, one or more than one of the answers may be correct. Select—

>   **A** if (1), (2), and (3) are correct;
>   **B** if (1) and (3) are correct;
>   **C** if (2) and (4) are correct;

D if only (4) is correct; and
E if all are correct.

8. TSH produces—
   (1) increased thyroidal uptake of iodine.
   (2) increased coupling of monoiodotyrosine and diiodotyrosine.
   (3) increased synthesis of thyroglobulin.
   (4) increased cyclic AMP levels in thyroid cells.

   A   B   C   D   E

9. T_4—
   (1) is formed by condensation of 2 iodinated tyrosine residues.
   (2) binds directly to DNA in the nucleus of target cells.
   (3) is bound to protein in the plasma.
   (4) increases plasma cholesterol levels.

   A   B   C   D   E

10. The following hormones have α-subunits with the same structure—
    (1) FSH.
    (2) human chorionic somatomammotropin (hCS).
    (3) TSH.
    (4) prolactin.

    A   B   C   D   E

11. Hypothyroidism due to disease of the thyroid gland is associated with increased plasma levels of—
    (1) RT_3.
    (2) cholesterol.
    (3) albumin.
    (4) TSH.

    A   B   C   D   E

12. Exophthalmos (infiltrative ophthalmopathy) is—
    (1) improved after thyroidectomy.
    (2) not improved by propylthiouracil or other thiocarbamide drugs.
    (3) a common complication of myxedema.
    (4) probably due to circulating antithyroid immunoglobulins that act on tissue in the orbit.

    A   B   C   D   E

13. Use of thyroid hormones to treat euthyroid patients—
    (1) is unwise because it increases the plasma cholesterol concentration.
    (2) is of value in relieving the symptoms of patients who are weak and tired but have a normal BMR.
    (3) is a simple and effective way to treat obesity.
    (4) produces no change in the BMR unless the dose is relatively large.

    A   B   C   D   E

14. T_3—
   (1) is derived from tryptophan.
   (2) is synthesized in the thyroid gland.
   (3) is less active than RT_3.
   (4) is synthesized in tissues outside the thyroid gland.

A   B   C   D   E

15. The thyroid hormone receptor—
   (1) binds T_4 as well as T_3 but with a lesser affinity.
   (2) is found in 2 forms encoded by different genes.
   (3) has a DNA-binding domain.
   (4) is homologous to the α_2 adrenergic receptor.

A   B   C   D   E

In the following questions, indicate whether the item on the left is greater than **(G)**, the same as **(S)**, or less than **(L)** the item on the right.

16. Plasma cholesterol level in myxedema.   G S L   Plasma cholesterol level in thyrotoxicosis.

17. Food intake in myxedema.   G S L   Food intake in thyrotoxicosis.

18. Number of β receptors in the heart in myxedema.   G S L   Number of β receptors in the heart in thyrotoxicosis.

19. Thyroidal uptake of radio-active iodine 4 hours after a single dose of propylthio-uracil.   G S L   Thyroidal uptake of radio-active iodine 4 hours after a single dose of perchlorate.

20. Thyroidal uptake of radio-active iodine 4 hours after starting propylthiouracil treatment.   G S L   Thyroidal uptake of radio-active iodine 24 hours after starting propylthiouracil treatment.

21. TSH secretion following injection of TRH in hypo-thyroidism due to disease of the thyroid gland.   G S L   TSH secretion following injection of TRH in hyper-thyroidism due to circulat-ing antibody with TSH activity.

22. Plasma free T_4 in a patient with a subnormal plasma total thyroid hormone con-centration due to a low plasma concentration of TBG.   G S L   Plasma free T_4 in a patient with an elevated plasma total thyroid hormone con-centration due to a high plasma concentration of TBG.

23. Plasma TSH level in an infant upon exposure to cold.   G S L   Plasma TSH level in an adult upon exposure to cold.

24. Plasma RT_3 concentration after 7 days of starvation.   G S L   Plasma RT_3 concentration after 14 days of starvation.

25. Amount of endocytosis at the colloid-thyroid cell border following injection of TSH.

G   S   L

Amount of endocytosis at the colloid-thyroid cell border following injection of FSH.

REFERENCES

Ingbar SH, Woeber KA: The thyroid gland. Chapter 4 in: *Textbook of Endocrinology,* 6th ed. Williams RH (editor). Saunders, 1981.

Norman AW, Litwack G: *Hormones* Academic Press, 1987.

Endocrine Functions of the Pancreas & the Regulation of Carbohydrate Metabolism

19

Chapter 19 is concerned primarily with insulin and glucagon, the 2 principal carbohydrate-regulating hormones secreted by the pancreas. Diabetes mellitus, the syndrome that results from insulin deficiency, is reviewed in detail, along with hypoglycemia. In addition, pancreatic polypeptide and somatostatin, the other hormones of the pancreatic islets, are discussed, and the many other hormones in the body that affect carbohydrate metabolism are reviewed.

OBJECTIVES

The material in the chapter should help students to—

- List the hormones that affect the blood glucose concentration, and briefly describe the action of each.

- Describe the structure of the pancreatic islets, and name the hormones secreted by each of the cell types in the islets.

- Describe the structure of insulin, and outline the steps involved in its biosynthesis and release into the bloodstream.

- List the consequences of insulin deficiency, and explain how each of these abnormalities is produced.

- Describe insulin receptors, the way they are regulated, and the way they mediate the effects of insulin.

- Describe the effects of insulin excess, and summarize the homeostatic mechanisms that combat hypoglycemia.

- List the major factors that affect the secretion of insulin.

- Describe the structure of glucagon and its relation to glicentin.

- List the physiologically significant effects of glucagon.

- List the principal factors that affect the secretion of glucagon.

- Describe the probable physiologic effects of somatostatin in the pancreas.

- Outline the mechanisms by which thyroid hormones, adrenal glucocorticoids, catecholamines, and growth hormone affect carbohydrate metabolism.

- List the common causes of hypoglycemia in humans, and describe how each acts to lower the blood sugar concentration.

- Define type I (insulin-dependent) and type II (non-insulin-dependent) diabetes mellitus, and describe the principal features and probable causes of each type.

GENERAL QUESTIONS

1. A patient has hypoglycemia. He may have an insulin-secreting tumor or functional hypoglycemia, but there is also reason to believe that he may be surreptitiously injecting himself with insulin. How would you differentiate among these possibilities? What tests would you order, and what would each tell you?

2. Which of the curves in Fig 19–1 would you expect to see in a patient with diabetes mellitus? Explain your answer.

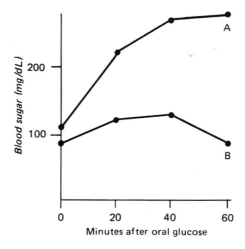

Figure 19–1.

3. Which of the curves in Fig 19–2 would you expect to see (1) in a normal individual, (2) in an individual with type I (insulin-dependent) diabetes mellitus, and (3) in an individual with type II (non-insulin-dependent) diabetes mellitus? Explain the differences in the curves.

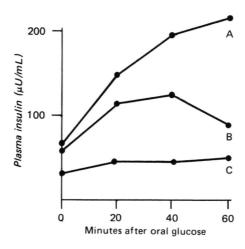

Figure 19–2.

4. Describe the distribution of the various types of cells within the pancreatic islets. Does the distribution differ in different parts of the pancreas?

5. Discuss the forms and causes of diabetes mellitus in humans.

6. Why is ketosis more common and more severe in type I diabetes than in type II diabetes?

7. It has been argued that the pancreatic islets are a site where hormones act in a paracrine fashion. Summarize and evaluate the evidence for this hypothesis.

8. Describe the changes in intermediary metabolism that occur in the first 3 hours after a meal containing large quantities of carbohydrate and protein. What roles do hormones play in these changes?

9. Discuss glucagon in terms of the biosynthesis of glucagon and glicentin from the preprohormone, the locations in which glucagon is found in the body, and the function of glucagon in each of these locations.

10. Insulin is sometimes described as "the hormone of plenty" and glucagon as "the hormone of starvation." Discuss the appropriateness of these terms in the context of the physiologic effects of each hormone and their interactions.

11. What is proinsulin? What is its relation to preproinsulin, and what is its fate?

12. Insulin acts on receptors in the membranes of many different cells to produce metabolic effects. Describe the structure of insulin receptors, the events triggered when insulin binds to these receptors, and the conditions that cause variations in the number and affinity of insulin receptors.

MULTIPLE-CHOICE QUESTIONS

In the following questions, select the single best answer.

1. Which of the following are *incorrectly* paired?
    a. B cells : insulin.
    b. D cells : somatostatin.
    c. A cells : glucagon.
    d. pancreatic exocrine cells : chymotrypsinogen.
    e. F cells : gastrin.

2. Which of the following would be *least* likely to occur several days after a rat is injected with a drug that kills all of its pancreatic B cells?
    a. a rise in the plasma H^+ concentration.
    b. a rise in the plasma glucagon concentration.
    c. a fall in the plasma HCO_3^- concentration.
    d. a fall in the plasma amino acid concentration.
    e. a rise in plasma osmolality.

3. Glucagon is *not* normally found in the—
    a. brain.
    b. pancreas.
    c. gastrointestinal tract.
    d. plasma.
    e. adrenal glands.

4. When the blood glucose concentration falls to low levels, a number of different hormones help combat the hypoglycemia. After intravenous administration of a large dose of insulin, the return of a low blood sugar level to normal is delayed in—
   a. adrenal medullary insufficiency.
   b. thyrotoxicosis.
   c. glucagon deficiency.
   d. combined adrenal medullary insufficiency and glucagon deficiency.
   e. diabetes mellitus.

5. Which of the following is *not* seen following total pancreatectomy?
   a. little change in the plasma insulin level.
   b. little change in the plasma glucagon level.
   c. steatorrhea.
   d. increased plasma levels of free fatty acids.
   e. decreased plasma P_{CO_2}.

6. Which of the following are *incorrectly* paired?
   a. epinephrine : increased glycogenolysis in skeletal muscle.
   b. insulin : increased protein synthesis.
   c. glucagon : increased gluconeogenesis.
   d. progesterone : increased plasma glucose level.
   e. growth hormone : increased plasma glucose level.

7. Insulin increases the entry of glucose into—
   a. all tissues.
   b. renal tubular cells.
   c. the mucosa of the small intestine.
   d. neurons in the cerebral cortex.
   e. skeletal muscle.

8. The mechanism by which glucagon produces an increase in the blood glucose concentration involves—
   a. binding of glucagon to DNA in the nucleus of target cells.
   b. binding of glucagon to receptors in the cytoplasm of target cells.
   c. activation of G_s in target cells.
   d. increased binding of Ca^{2+} in the cytoplasm of target cells.
   e. inhibition of insulin secretion.

9. Glucagon increases glycogenolysis in liver cells but ACTH does not because—
   a. cortisol increases the plasma glucose level.
   b. liver cells have an adenylate cyclase different from that in adrenocortical cells.
   c. ACTH cannot enter the nucleus of liver cells.
   d. the membranes of liver cells contain receptors that bind glucagon but lack receptors that bind ACTH.
   e. liver cells contain a protein that inhibits the action of ACTH.

10. A meal rich in proteins containing the amino acids that stimulate insulin secretion but low in carbohydrate does not cause hypoglycemia, because—
   a. the meal causes a compensatory increase in T_4 secretion.
   b. cortisol in the circulation prevents glucose from entering muscle.
   c. glucagon secretion is also stimulated by the meal.
   d. the amino acids in the meal are promptly converted to glucose.
   e. insulin does not decrease the blood glucose level if the plasma concentration of amino acids is elevated.

In the following questions, one or more than one of the answers may be correct.
Select—

>**A** if (1), (2), and (3) are correct;
>**B** if (1) and (3) are correct;
>**C** if (2) and (4) are correct;
>**D** if only (4) is correct; and
>**E** if all are correct.

11. Which of the following stimulate the secretion of *both* insulin and glucagon?
(1) acetylcholine.
(2) β-adrenergic agonists.
(3) certain amino acids.
(4) glucose.

>**A** **B** **C** **D** **E**

12. Insulin secretion is increased by—
(1) glucose.
(2) amino acids.
(3) gastric inhibitory peptide (GIP).
(4) sulfonylureas.

>**A** **B** **C** **D** **E**

13. Which of the following hormones are normally secreted in adult humans by cells in the pancreatic islets?
(1) gastrin.
(2) pancreatic polypeptide.
(3) GRH.
(4) somatostatin.

>**A** **B** **C** **D** **E**

14. Which of the following are frequently found in patients with type II (non-insulin-dependent) diabetes?
(1) microangiopathy.
(2) obesity.
(3) arteriosclerosis.
(4) severe ketosis.

>**A** **B** **C** **D** **E**

15. A patient with a blood sugar level of 30 mg/dL probably has elevated plasma concentrations of—
(1) glucagon.
(2) epinephrine.
(3) cortisol.
(4) growth hormone.

>**A** **B** **C** **D** **E**

16. Insulin produces—
    (1) increased deposition of fat.
    (2) increased entry of amino acids into cells.
    (3) increased entry of K^+ into muscle cells.
    (4) increased insertion of glucose transporters into cell membranes.

A   **B**   **C**   **D**   **E**

17. Insulin secretion is affected by—
    (1) the sympathetic nervous system.
    (2) the level of glucose in the blood.
    (3) the level of amino acids in the blood.
    (4) the parasympathetic nervous system.

A   **B**   **C**   **D**   **E**

18. Large doses of glucagon—
    (1) increase the force of contraction of the heart.
    (2) increase the concentration of amino acids in plasma.
    (3) increase the concentration of free fatty acids in plasma.
    (4) decrease the plasma Na^+ concentration.

A   **B**   **C**   **D**   **E**

19. Physical stress such as breaking an arm increases the availability of glucose to the brain because—
    (1) the plasma cortisol level rises.
    (2) epinephrine secretion is increased.
    (3) less glucose enters skeletal muscles.
    (4) T_4 secretion is increased.

A   **B**   **C**   **D**   **E**

20. In a patient with untreated severe diabetes mellitus, you would expect to find—
    (1) an increase in the rate and depth of breathing.
    (2) elevated plasma free fatty acid levels.
    (3) the odor of acetone on the breath.
    (4) increased plasma H^+ concentration.

A   **B**   **C**   **D**   **E**

In the following questions, indicate whether the item on the left is greater than **(G)**, the same as **(S)**, or less than **(L)** the item on the right.

21. Plasma glucagon level after eating candy.   **G**   **S**   **L**   Plasma glucagon level after eating meat.

22. Insulin resistance in Cushing's syndrome (hypercortisolism).   **G**   **S**   **L**   Insulin resistance in adrenocorticoid insufficiency (Addison's disease).

23. Blood sugar level after surgical pancreatectomy.   **G**   **S**   **L**   Blood sugar level after administration of a toxin such as alloxan that selectively kills B cells.

24. Blood sugar level 2 hours after a meal in hyperthyroidism.   **G S L**   Blood sugar level 2 hours after a meal in hypothyroidism.

25. Insulin secretion after glucose is given intravenously.   **G S L**   Insulin secretion after the same amount of glucose is given orally.

26. Plasma ketone concentration 2 hours after a meal.   **G S L**   Plasma ketone concentration 24 hours after a meal.

27. Blood sugar level in a patient with an untreated pancreatic tumor secreting somatostatin.   **G S L**   Blood sugar level in a patient with an untreated pancreatic tumor secreting gastrin.

28. Plasma concentration of C peptide after eating carbohydrate.   **G S L**   Plasma concentration of C peptide after eating fat.

29. Rise in plasma insulin level after infusion of epinephrine.   **G S L**   Rise in plasma glucagon level after infusion of the same dose of epinephrine.

30. Target cell content of cyclic AMP after administration of glucagon.   **G S L**   Target cell content of cyclic AMP after administration of insulin.

REFERENCES

Tepperman J: *Metabolic and Endocrine Physiology,* 4th ed. Chapter 14: Endocrine function of the pancreas. Year Book, 1980.

20

The Adrenal Medulla & Adrenal Cortex

Chapter 20 is concerned with the multiple functions of the adrenal glands. The catecholamines secreted by the adrenal medulla are considered first, and the mechanisms regulating their secretion are analyzed. The glucocorticoids and the hormones secreted by the inner 2 zones of the adrenal cortex are reviewed next, along with their control by ACTH. The mineralocorticoid aldosterone, which is secreted by the outer zone of the adrenal cortex, is also reviewed, and the role of angiotensin II, electrolytes, and ACTH in the regulation of aldosterone secretion is analyzed. Finally, there is a brief review of the clinical features of adrenocortical deficiency, Cushing's syndrome, Conn's syndrome, and the adrenogenital syndrome.

OBJECTIVES

The material in the chapter should help students to—

- Name the catecholamines secreted by the adrenal medulla, and outline their biosynthesis.

- Compare the actions of the 3 main catecholamines in the body.

- Name the principal metabolites of epinephrine and norepinephrine.

- List the principal stimuli that increase adrenal medullary secretion, and describe the way they bring about the increases.

- Differentiate between C_{19} and C_{21} steroids, and give examples of each.

- Outline the steps involved in steroid biosynthesis in the adrenal cortex.

- Name the plasma proteins that bind adrenocortical steroids, and discuss the role they play in adrenocortical physiology.

- Name the major site of adrenocortical hormone metabolism and the principal metabolites produced from glucocorticoids and aldosterone.

- List and briefly describe the physiologic and pharmacologic effects of glucocorticoids.

- Contrast the physiologic and pathologic effects of adrenal androgens.

- Describe the mechanisms that regulate secretion of glucocorticoids and adrenal sex hormones.

- Outline the actions of aldosterone.

- Describe the mechanisms that regulate aldosterone secretion.

- Describe the main features of the diseases caused by excess or deficiency of each of the hormones of the adrenal gland.

GENERAL QUESTIONS

1. What are the principal hormones secreted by each of the 3 zones of the adrenal cortex? What enzymes are unique to each zone, ie, found only in that particular zone?

2. What are the functions of the human fetal adrenal cortex?

3. Which of the metabolites of norepinephrine and epinephrine would you measure to determine whether a patient with pheochromocytoma had a norepinephrine- or epinephrine-secreting tumor?

4. What would you expect to be the consequences of 17α-hydroxylase deficiency, an inherited genetic disorder? Explain your answer.

5. Why do some patients with congenital adrenal hyperplasia and virilization have high blood pressure but others have excessive loss of sodium and hypotension? Explain your answer.

6. Patients with nephrosis sometimes have low plasma 17-hydroxycorticoid levels but do not develop signs and symptoms of adrenal insufficiency. Why?

7. In patients with liver disease, glucocorticoids are metabolized at a lower rate than normal, yet signs of glucocorticoid excess do not develop. Why?

8. List the possible causes of Cushing's syndrome. How would you treat each of them?

9. What are the advantages and disadvantages of long-term, high-dose treatment with glucocorticoids in diseases such as rheumatoid arthritis and asthma?

10. After hypophysectomy, the responsiveness of the adrenal cortex to ACTH is reduced, and after ACTH treatment, the response to subsequent doses of ACTH is increased. Explain these alterations in adrenal responsiveness.

11. When first administered, ACTH increases aldosterone secretion, but with continuing treatment aldosterone secretion subsequently declines to low levels. Why does this biphasic response occur?

12. Describe how the renin-angiotensin-aldosterone system operates to maintain extracellular fluid volume at a normal level.

MULTIPLE-CHOICE QUESTIONS

In the following questions, select the single best answer.

1. Which of the following is *not* produced by physiologic amounts of glucocorticoids?
   a. normal responsiveness of fat depots to norepinephrine.
   b. maintenance of normal vascular reactivity.
   c. increased excretion of a water load.
   d. inhibition of the inflammatory response.
   e. inhibition of ACTH secretion.

2. The zona fasciculata of the adrenal cortex produces cortisol but the zona glomerulosa does not, because—
   a. the zona fasciculata lacks 11β-hydroxylase activity but the zona glomerulosa possesses it.

    b. the zona fasciculata cannot make as much deoxycorticosterone as the zona glomerulosa can.

    c. the zona fasciculata has 17α-hydroxylase activity but the zona glomerulosa does not.

    d. the zona fasciculata has receptors for ACTH but the zona glomerulosa does not.

    e. the zona fasciculata has receptors for angiotensin II but the zona glomerulosa does not.

3. Which of the following reactions is the primary site of action of ACTH?

    a. progesterone→corticosterone.

    b. corticosterone→aldosterone.

    c. 17-hydroxypregnenolone→dehydroepiandrosterone.

    d. pregnenolone→17-hydroxypregnenolone.

    e. cholesterol→pregnenolone.

4. Which of the following reactions is blocked in congenital 17α-hydroxylase deficiency?

    a. progesterone→corticosterone.

    b. corticosterone→aldosterone.

    c. 17-hydroxypregnenolone→17-hydroxyprogesterone.

    d. pregnenolone→17-hydroxypregnenolone.

    e. cholesterol→pregnenolone.

5. Which of the following are *incorrectly* paired?

    a. gluconeogenesis : cortisol.

    b. free fatty acid mobilization : dehydroepiandrosterone.

    c. muscle glycogenolysis : epinephrine.

    d. kaliuresis : aldosterone.

    e. hepatic glycogenesis : insulin.

6. Which of the following hormones has the shortest half-life (is most rapidly destroyed in the blood)?

    a. ACTH.

    b. renin.

    c. dehydroepiandrosterone.

    d. aldosterone.

    e. norepinephrine.

7. Which of the following reactions produces a 17-ketosteroid?

    a. progesterone→corticosterone.

    b. corticosterone→aldosterone.

    c. 17-hydroxypregnenolone→dehydroepiandrosterone.

    d. pregnenolone→17-hydroxypregnenolone.

    e. cholesterol→pregnenolone.

8. Which of the following reactions is facilitated by angiotensin II but not by ACTH?

    a. progesterone→corticosterone.

    b. corticosterone→aldosterone.

    c. 17-hydroxypregnenolone→dehydroxypregnenolone.

    d. pregnenolone→17-hydroxypregnenolone.

    e. cholesterol→pregnenolone.

9. Mole for mole, which of the following has the greatest effect on Na^+ excretion?

    a. progesterone.

    b. cortisol.

    c. vasopressin.

    d. aldosterone.

    e. dehydroepiandrosterone.

10. Mole for mole, which of the following has the greatest effect on plasma osmolality?
    a. progesterone.
    b. cortisol.
    c. vasopressin.
    d. aldosterone.
    e. dehydroepiandrosterone.

In the following questions, one or more than one of the answers may be correct. Select—

> **A** if (1), (2), and (3) are correct;
> **B** if (1) and (3) are correct;
> **C** if (2) and (4) are correct;
> **D** if only (4) is correct; and
> **E** if all are correct.

11. Transcortin—
    (1) is produced in the liver.
    (2) helps maintain a constant blood level of cortisol.
    (3) increases the half-life of cortisol in the blood.
    (4) increases the rate of filtration of cortisol in the glomeruli of the kidney.

    A   B   C   D   E

12. Cortisol increases blood glucose by—
    (1) decreasing glucose utilization in skeletal muscle.
    (2) increasing growth hormone secretion.
    (3) increasing gluconeogenesis.
    (4) increasing protein synthesis in muscle.

    A   B   C   D   E

13. Glucocorticoid secretion would be expected to decrease following injection of a drug that—
    (1) inhibits 21β-hydroxylase in the adrenal gland.
    (2) blocks conversion of cholesterol to pregnenolone.
    (3) prevents 11β-hydroxylation in the adrenal gland.
    (4) prevents conversion of 17-hydroxypregnenolone to dehydroepiandrosterone.

    A   B   C   D   E

14. Which of the following are *correctly* paired?
    (1) epinephrine : precursor for norepinephrine.
    (2) tyrosine : precursor for dopamine.
    (3) cortisone : precursor for cortisol.
    (4) androstenedione : precursor for estrogen.

    A   B   C   D   E

15. Epinephrine and norepinephrine—
    (1) are amino acids.
    (2) are both secreted by neurons in the autonomic nervous system.
    (3) are polypeptides.
    (4) both activate α- and β-adrenergic receptors.

    A   B   C   D   E

16. An increase in which of the following increases aldosterone secretion by a direct action on the adrenal cortex?
(1) ACTH.
(2) renin.
(3) angiotensin II.
(4) sodium.

A   B   C   D   E

17. Which of the following hormones are synthesized as part of a much larger molecule?
(1) dopamine.
(2) ACTH.
(3) cortisol.
(4) angiotensin II.

A   B   C   D   E

18. A decrease in extracellular fluid volume would be expected to cause increased secretion of—
(1) vasopressin.
(2) renin.
(3) ACTH.
(4) dehydroepiandrosterone.

A   B   C   D   E

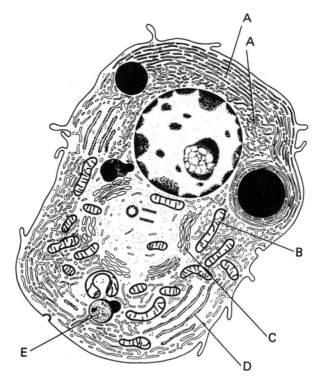

Figure 20–1. Steroid-secreting cell. (Reproduced, with permission, from Fawcett DW, Long JA, Jones AL: The ultrastructure of endocrine glands. *Recent Prog Horm Res* 1969;25:315.)

In the following questions, select the letter designating the part of the steroid-secreting cell in Fig 20–1 that manufactures or contains the material. A lettered part may be selected once, more than once, or not at all.

19.	Acid phosphatase	**A**	**B**	**C**	**D**	**E**
20.	Pregnenolone	**A**	**B**	**C**	**D**	**E**
21.	Progesterone	**A**	**B**	**C**	**D**	**E**
22.	ACTH receptors	**A**	**B**	**C**	**D**	**E**
23.	Glycoproteins	**A**	**B**	**C**	**D**	**E**

In the following questions, indicate whether the item on the left is greater than **(G)**, the same as **(S)**, or less than **(L)** the item on the right.

24. Plasma concentration of ACTH in a human at 4:00 AM.   **G S L**   Plasma concentration of ACTH in a human at 4:00 PM.

25. 18-Hydroxylase concentration in the zona glomerulosa.   **G S L**   18-Hydroxylase concentration in the zona fasciculata.

26. Blood glucose level in a patient with an epinephrine-secreting pheochromocytoma.   **G S L**   Blood glucose level in a patient with a norepinephrine-secreting pheochromocytoma.

27. Diastolic blood pressure in a patient with an epinephrine-secreting pheochromocytoma.   **G S L**   Diastolic blood pressure in a patient with a norepinephrine-secreting pheochromocytoma.

28. Heart rate in a patient with an epinephrine-secreting pheochromocytoma.   **G S L**   Heart rate in a patient with a norepinephrine-secreting pheochromocytoma.

29. Urinary excretion of aldosterone when sodium intake is 15 meq/d.   **G S L**   Urinary excretion of aldosterone when sodium intake is 75 meq/d.

30. Blood pressure in a patient with congenital 11β-hydroxylase deficiency.   **G S L**   Blood pressure in a patient with congenital 21β-hydroxylase deficiency.

31. Incidence of osteoporosis in adrenogenital syndrome.   **G S L**   Incidence of osteoporosis in Cushing's syndrome.

32. Skin pigmentation in a patient with adrenal insufficiency due to adrenal disease.   **G S L**   Skin pigmentation in a patient with adrenal insufficiency due to pituitary disease.

33. Na$^+$-K$^+$ ATPase in cell
    membrane of cell stimu-
    lated by aldosterone.

G S L   Na$^+$-K$^+$ ATPase in cell
               membrane of cell stimu-
               lated by cortisol.

REFERENCES

Martin CR: *Endocrine Physiology*. Chaps 6–7, pp 215–269, and Chap 9, pp 321–368. Oxford Univ Press, 1985.

Ungar A, Phillips JH: Regulation of the adrenal medulla. *Physiol Rev* 1983;**63:**787.

Hormonal Control of Calcium Metabolism & the Physiology of Bone

21

In Chapter 21, the metabolism of calcium and phosphorus is discussed and the physiology of bone is reviewed. The 3 hormones that are the primary regulators of calcium metabolism—1,25-dihydroxycholecalciferol, parathyroid hormone, and calcitonin—are discussed in detail. In addition, there is a brief review of the effects of other hormones on calcium metabolism.

OBJECTIVES

The material in the chapter should help students to—

- Describe the distribution of calcium in the body and the forms in which it exists in plasma.

- Name the factors affecting plasma Ca^{2+} concentration, and discuss the mechanism by which each exerts its effects.

- Summarize the distribution of phosphorus in the body.

- Name the types of cells found in bone, and describe the function of each.

- Outline the ways that bones grow.

- Describe the formation of vitamin D in the skin, its subsequent hydroxylation in the liver and kidneys, and the actions of its biologically active metabolites.

- Describe the biosynthesis and metabolism of parathyroid hormone.

- Describe the actions of parathyroid hormone.

- Identify the source of calcitonin, its chemical nature, and its principal actions.

- Summarize the effects of glucocorticoids, growth hormone, and growth factors on Ca^{2+} metabolism.

GENERAL QUESTIONS

1. How would you demonstrate that parathyroid hormone is produced in and secreted by the parathyroid glands? How would you identify the cells in which it is produced?

2. Name the tissues in the body on which 1,25-dihydroxycholecalciferol, parathyroid hormone, and calcitonin act, and describe the effects of these hormones on each of these tissues.

3. Normally, the total amount of Ca^{2+} in the extracellular fluid is about 22.5 mmol. How is Ca^{2+} added to the extracellular fluid, and how is it removed?

4. What are the effects of hypocalcemia and hypercalcemia?

5. What are osteoprogenitor cells? How do they relate to the other types of cells in bone?

6. Discuss the physiology of the epiphyseal plate. What is epiphyseal closure, and what produces it?

7. What are the signs and symptoms of hyperparathyroidism, and what causes each of them?

8. Define osteomalacia, osteosclerosis, and osteoporosis. Discuss the causes and treatment of osteoporosis.

9. Discuss calcification, considering both the bone matrix and the calcification process itself.

10. There was a sharp increase in the incidence of bone disease with the onset of the Industrial Revolution in the 19th century, but the incidence has since decreased. What bone disease was produced, and why? Why has the incidence declined in the 20th century?

11. Discuss renal 1α-hydroxylase and its regulation. Why is this enzyme of great importance in the production of healthy bone?

12. On the axes in Fig 21–1, plot the relationship of plasma calcitonin concentration to plasma Ca^{2+} concentration and that of plasma parathyroid hormone concentration to plasma Ca^{2+} concentration.

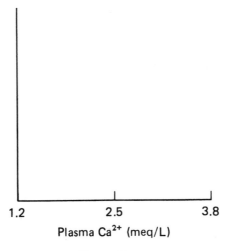

Plasma Ca^{2+} (meq/L)

Figure 21–1.

MULTIPLE-CHOICE QUESTIONS

In the following questions, select the single best answer.

1. A patient with parathyroid deficiency 10 days after thyroidectomy with damage to the parathyroid glands would probably have—
   a. low plasma phosphate and Ca^{2+} levels and tetany.
   b. low plasma phosphate and Ca^{2+} levels and tetanus.

    c. a low plasma Ca^{2+} level, increased muscular excitability, and a characteristic spasm of the muscles of the upper extremity (Trousseau's sign).

    d. high plasma phosphate and Ca^{2+} levels and bone demineralization.

    e. increased muscular excitability, a high plasma Ca^{2+} level, and bone demineralization.

2. A high plasma Ca^{2+} level causes—
    a. bone demineralization.
    b. increased formation of 1,25-dihydroxycholecalciferol.
    c. decreased secretion of calcitonin.
    d. decreased blood coagulability.
    e. increased formation of 24,25-dihydroxycholecalciferol.

3. Which of the following is *not* involved in regulating plasma Ca^{2+} levels?
    a. the kidneys.
    b. the skin.
    c. the liver.
    d. the lungs.
    e. the intestine.

4. Which of the following exerts the greatest effect on parathyroid hormone secretion?
    a. the plasma phosphate concentration.
    b. calcitonin.
    c. 1,25-dihydroxycholecalciferol.
    d. the total plasma calcium concentration.
    e. the plasma Ca^{2+} concentration.

5. 1,25-Dihydroxycholecalciferol affects intestinal Ca^{2+} absorption through a mechanism that—
    a. is comparable to the action of steroid hormones.
    b. activates adenylate cyclase.
    c. increases cell turnover.
    d. changes gastric acid secretion.
    e. is comparable to the action of polypeptide hormones.

In the following questions, one or more than one of the answers may be correct. Select—

                A if (1), (2), and (3) are correct;
                B if (1) and (3) are correct;
                C if (2) and (4) are correct;
                D if only (4) is correct; and
                E if all are correct.

6. Ca^{2+} is important in many different biologic processes. It plays an important role in—
    (1) contraction of cardiac muscle.
    (2) release of transmitters and hormones from neurons and gland cells.
    (3) contraction of skeletal muscle.
    (4) blood coagulation.

             A   **B**   **C**   **D**   **E**

7. Which of the following would you expect to find in a patient whose diet has been low in calcium for 2 months?
    (1) decreased formation of 24,25-dihydroxycholecalciferol.
    (2) increased amounts of calcium-binding protein in intestinal epithelial cells.

(3) increased parathyroid hormone secretion.
(4) a high plasma calcitonin concentration.

A   **B**   **C**   **D**   **E**

8. Which of the following would tend to *increase* the plasma Ca^{2+} level?
    (1) increased intestinal absorption of Ca^{2+}.
    (2) a low plasma protein level.
    (3) a low plasma pH.
    (4) decreased phosphate reabsorption in the renal tubules.

A   **B**   **C**   **D**   **E**

9. Epiphyseal width is regulated by—
    (1) calcitonin.
    (2) estrogens.
    (3) glucocorticoids.
    (4) somatomedins.

A   **B**   **C**   **D**   **E**

10. Calcium metabolism is affected in—
     (1) pseudohypoparathyroidism.
     (2) renal failure.
     (3) cancer.
     (4) precocious puberty.

A   **B**   **C**   **D**   **E**

In the following questions, indicate whether the item on the left is greater than **(G)**, the same as **(S)**, or less than **(L)** the item on the right.

11. Percent of ingested Ca^{2+} absorbed from the intestine in a subject fed a high-calcium diet.   **G S L**   Percent of ingested Ca^{2+} absorbed from the intestine in a subject fed a low-calcium diet.

12. Renal reabsorption of phosphate in a subject fed a high-calcium diet.   **G S L**   Renal reabsorption of phosphate in a subject fed a low-calcium diet.

13. Plasma calcitonin concentration after drinking milk.   **G S L**   Plasma calcitonin concentration after drinking chicken broth.

14. Plasma parathyroid hormone concentration in pseudohypoparathyroidism.   **G S L**   Plasma parathyroid hormone concentration in hypoparathyroidism.

15. Total amount of calcium in the skeleton of a 20-year-old woman.   **G S L**   Total amount of calcium in the skeleton of a 60-year-old woman.

16. Incidence of osteoporosis in men.   **G   S   L**   Incidence of osteoporosis in women.

REFERENCES

Fregly MJ, Luttge WG: *Human Endocrinology: An Interactive Text.* Chap 10, pp 297–317. Elsevier, 1982.

Hershman JM (editor): *Endocrine Pathophysiology,* 2nd ed. Chap 9, pp 228–253. Lea & Febiger, 1982.

22   The Pituitary Gland

Chapter 22 reviews the formation of the pituitary gland in the fetus and its morphology in the adult. The intermediate lobe hormones and their physiology in various species are discussed, as well as the structure and breakdown of the large hormone precursor molecule pro-opiomelanocortin, which is synthesized and processed in the anterior and intermediate lobes of the gland. The posterior pituitary hormones are discussed in Chapter 14, and most of the anterior pituitary hormones are considered in the chapters on the glands they regulate, but growth hormone is discussed in detail in this chapter. IGF-I and IGF-II, the principal somatomedins, are also reviewed.

OBJECTIVES

The material in the chapter should help students to—

- Name the hormones secreted by the pituitary gland in humans, and list the main functions of each.

- Describe the 3 lobes of the pituitary gland in terms of their embryologic origin and their structure in adults.

- Name the 5 main types of secretory cells found in the anterior pituitary, and note whether each is acidophilic, basophilic, or chromophobic.

- Describe the mechanisms responsible for changes in skin coloration in species in which such changes take place, and name the hormones involved.

- Describe the pro-opiomelanocortin molecule and the products produced from it in the anterior and intermediate lobes of the pituitary gland.

- Describe the structure of growth hormone, and list its actions.

- Describe the relation of IGF-I and IGF-II to the actions of growth hormone.

- List the 2 hypothalamic hypophyseotropic hormones that regulate growth hormone secretion and the principal stimuli that bring about increases or decreases in growth hormone secretion.

- List the factors needed for normal growth, and describe the contribution each makes during prenatal and postnatal development.

- Describe and explain the main features of hypopituitarism, acromegaly, and Nelson's syndrome.

GENERAL QUESTIONS

1. What endocrine diseases cause alterations in skin pigmentation in humans? What hormones are involved?

2. Although the same pro-opiomelanocortin molecule is found in the anterior

and intermediate lobes of the pituitary gland and in neurons in the brain, different products are formed from it in each location. How can different products be formed from the same molecule? What processes are involved?

3. The structure of growth hormone varies in different animal species. How do these variations relate to differences in activity when growth hormone from one species is injected into another species?

4. What is sulfation factor?

5. How would you decide whether an increase in growth hormone secretion was due to decreased secretion of somatostatin or increased secretion of GRH?

6. Why does destruction of the posterior lobe of the pituitary gland cause marked polyuria whereas destruction of the whole pituitary gland causes little if any polyuria?

7. Discuss the relationship of nutrition to growth.

8. Discuss the endocrinologic causes of dwarfism, and explain how each leads to short stature.

9. How is human growth hormone (hGH) related to human chorionic somato-mammotropin (hCS)?

10. Why does excess growth hormone secretion in adults not cause increased stature?

11. In what order would you expect the functions of the anterior pituitary gland to disappear in a patient with a nonsecreting tumor that gradually destroys the gland? Explain your answer.

MULTIPLE-CHOICE QUESTIONS

In the following questions, select the single best answer.

1. Which of the following hormones exerts the least effect on growth?
   a. growth hormone.
   b. testosterone.
   c. T_4.
   d. insulin.
   e. vasopressin.

2. Which of the following are *incorrectly* paired?
   a. intermediate lobe : corticotropinlike intermediate lobe peptide (CLIP).
   b. anterior lobe : growth hormone.
   c. Laron dwarfism : low plasma IGF-I concentration.
   d. African pygmies : high plasma IGF-I concentration.
   e. acromegaly : high plasma IGF-I concentration.

3. Which of the following pituitary hormones is an opioid peptide?
   a. α-melanocyte-stimulating hormone (α-MSH).
   b. β-MSH.
   c. ACTH.
   d. β-endorphin.
   e. growth hormone.

4. Which of the following hormones is *not* made up of α- and β-subunits?
   a. TSH.
   b. luteinizing hormone (LH).
   c. follicle-stimulating hormone (FSH).
   d. human chorionic gonadotropin (hCG).
   e. prolactin.

In the following questions, one or more than one of the answers may be correct. Select—

    A if (1), (2), and (3) are correct;
    B if (1) and (3) are correct;
    C if (2) and (4) are correct;
    D if only (4) is correct; and
    E if all are correct.

      A   **B**   **C**   **D**   **E**

5. Hypopituitarism is characterized by—
   (1) infertility.
   (2) pallor.
   (3) a low basal metabolic rate.
   (4) intolerance to heat.

      A   **B**   **C**   **D**   **E**

6. The effects of IGF-I include—
   (1) stimulation of protein synthesis.
   (2) stimulation of growth.
   (3) lowering of blood glucose levels.
   (4) stimulation of insulin secretion.

      A   **B**   **C**   **D**   **E**

7. Which of the following can be found both in endocrine glands and in the brain?
   (1) somatostatin.
   (2) β-endorphin.
   (3) dopamine.
   (4) ACTH.

      A   **B**   **C**   **D**   **E**

8. In humans, growth hormone secretion is increased by—
   (1) arginine and certain other amino acids.
   (2) L-dopa.
   (3) stressful stimuli.
   (4) glucagon.

      A   **B**   **C**   **D**   **E**

9. Growth hormone excess causes—
   (1) acromegaly in adults.
   (2) gigantism in children.

(3) exhaustion atrophy of the pancreatic B cells when the growth hormone excess has been present for a long time.

(4) increased entry of glucose into muscle.

A   B   C   D   E

10. A scientist finds that infusion of growth hormone into the median eminence of the hypothalamus in experimental animals inhibits the secretion of growth hormone. Do you accept his conclusion that this proves that growth hormone feeds back to inhibit GRH secretion?

(1) no, because substances placed in the median eminence are easily transported to the anterior pituitary.

(2) no, because growth hormone does not cross the blood-brain barrier.

(3) no, because the infused growth hormone could be stimulating somatostatin secretion.

(4) yes, because systemically administered growth hormone inhibits growth hormone secretion.

A   B   C   D   E

In the following questions, indicate whether the item on the left is greater than **(G)**, the same as **(S)**, or less than **(L)** the item on the right.

11. Plasma growth hormone level when the blood sugar level is 95 mg/dL.	**G S L**	Plasma growth hormone level when the blood sugar level is 35 mg/dL.
12. Plasma growth hormone level after 30 minutes of sleep.	**G S L**	Plasma growth hormone level after 8 hours of sleep.
13. Plasma growth hormone level in a 5-year-old child.	**G S L**	Plasma growth hormone level in a 25-year-old adult.
14. Age at the time of pubertal growth spurt in girls.	**G S L**	Age at the time of pubertal growth spurt in boys.
15. Insulin sensitivity in hypopituitarism.	**G S L**	Insulin sensitivity in acromegaly.
16. *Rate* of growth in 5-year-old boy.	**G S L**	*Rate* of growth in 15-year-old boy.

REFERENCES

Daughaday WH (editor): *Endocrine Control of Growth.* Elsevier, 1982.

23

The Gonads: Development & Function of the Reproductive System

Chapter 23 is a review of the physiology of the reproductive system in adult males and females. The physiology of lactation and pregnancy, sexual differentiation and development in the fetus, puberty, and the menopause are also discussed.

OBJECTIVES

The material in the chapter should help students to—

- Name the important hormones secreted by the Leydig cells and Sertoli cells of the testes and by the graafian follicles and corpora lutea of the ovaries.

- Outline the role of chromosomes, hormones, and related factors in sex determination and development.

- List the major chromosomal and hormonal abnormalities that cause genital abnormalities.

- Summarize the hormonal and other changes that occur at puberty in males and females.

- Outline the hormonal and other changes that occur at menopause.

- Describe the hypothalamic mechanisms that regulate prolactin secretion, and list the principal stimuli that increase prolactin secretion and the drugs that decrease prolactin secretion.

- Outline the steps involved in spermatogenesis, from the primitive germ cells to mature, motile spermatozoa.

- List the principal components of semen.

- Describe the mechanisms that produce erection and ejaculation.

- Know the general structure of testosterone, and discuss its biosynthesis, transport, and metabolism.

- List the actions of testosterone.

- Outline the mechanisms involved in regulation of testosterone secretion.

- Describe the changes that occur in the ovaries, uterus, cervix, and vagina during the menstrual cycle.

- Know the general structures of 17β-estradiol and progesterone, and describe their biosynthesis, transport, and metabolism.

- List the actions of estrogens.

- List the actions of progesterone.

- Describe the roles of the pituitary and the hypothalamus in the regulation of ovarian function, and the role of feedback loops in this process.

- Describe the hormonal changes that accompany pregnancy.

- Comment on the role of hormones in parturition.

- Outline the processes involved in development of the breasts, production of milk, milk ejection, and termination of lactation.

GENERAL QUESTIONS

1. Describe the structure and function of the blood-testis barrier.

2. What are the effects on subsequent sexual development when a human female fetus is exposed to excess androgens in utero?

3. What are the effects on subsequent sexual development when normal testes fail to develop in a human male fetus?

4. What adult structures develop from the müllerian and wolffian ducts in the fetus?

5. What is the effect of hypophysectomy on the uterus? What other operation would produce the same results?

6. What is the function of the oviducts? How do they perform this function?

7. What physiologic changes occur in females during sexual intercourse?

8. What is the function of the acrosome?

9. Discuss the processes responsible for converting sperm in the seminiferous tubules into fully mature, motile sperm capable of producing fertlization.

10. Using the axes in Fig 23–1, diagram the changes in plasma LH, FSH, estrogen, and progesterone occurring during the human menstrual cycle.

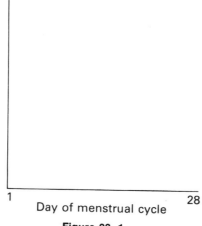

Day of menstrual cycle

Figure 23–1.

11. Draw pictures of the changes that occur in an ovary during the menstrual cycle, showing the important features of each stage from immature follicle to regressing corpus luteum.

12. Describe the testicular feminizing syndrome, and discuss its possible causes.

13. What is the evidence that the onset of puberty is under neural control?

14. Define eunuchoidism, and describe its characteristic features.

15. Compare and contrast the effects of castration in males before puberty with those of castration after puberty.

16. Using your knowledge of reproductive physiology, list the possible ways that contraception could be produced. How many of these are actually being used by the general public? Comment on the strengths and weaknesses of the 5 methods that you feel are most commonly used.

17. Using the axes in Fig 23–2, diagram the changes in plasma hCG, hCS, prolactin, growth hormone, and estradiol occurring during human pregnancy.

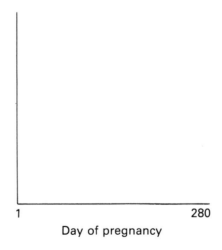

Day of pregnancy

Figure 23–2.

18. Discuss the relationship of hormones to cancer.

19. What is the fetoplacental unit, and what are its functions in terms of hormone production?

20. How would you decide whether amenorrhea in a young woman is due to pregnancy?

MULTIPLE-CHOICE QUESTIONS

In the following questions, select the single best answer.

1. If a young woman has high plasma levels of T_3, cortisol, and renin activity but her blood pressure is only slightly elevated and she has no symptoms

or signs of thyrotoxicosis or Cushing's syndrome, the most likely explanation is that—
- **a.** she has been treated with TSH and ACTH.
- **b.** she has been treated with T_3 and cortisol.
- **c.** she is in the third trimester of pregnancy.
- **d.** she has an adrenocortical tumor.
- **e.** she has been subjected to chronic stress.

2. Full development and function of the seminiferous tubules requires—
- **a.** somatostatin.
- **b.** LH.
- **c.** oxytocin.
- **d.** FSH.
- **e.** androgens and FSH.

3. In humans, fertilization usually occurs in the—
- **a.** vagina.
- **b.** cervix.
- **c.** uterine cavity.
- **d.** uterine tubes.
- **e.** abdominal cavity.

4. In human males, testosterone is produced mainly by the—
- **a.** Leydig cells.
- **b.** Sertoli cells.
- **c.** seminiferous tubules.
- **d.** epididymis.
- **e.** vas deferens.

5. The function of the seminiferous tubules is production of—
- **a.** immotile sperm.
- **b.** androgens.
- **c.** progesterone.
- **d.** motile sperm.
- **e.** renin.

6. Home-use kits for determining a woman's fertile period depend on the detection of one hormone in the urine. This hormone is—
- **a.** FSH.
- **b.** progesterone.
- **c.** estradiol.
- **d.** hCG.
- **e.** LH.

7. Puberty does not normally occur in humans under the age of 8 years, because before that age—
- **a.** the tissues are unresponsive to gonadal steroids.
- **b.** the ovaries and testes are unresponsive to gonadotropins.
- **c.** the pituitary cannot manufacture adequate amounts of gonadotropins.
- **d.** the brain secretes a substance that inhibits the responsiveness of the gonads to gonadotropins.
- **e.** the hypothalamus fails to secrete LHRH in a normal fashion.

8. Castration cells are found in the—
- **a.** uterus.
- **b.** prostate.
- **c.** testes.
- **d.** placenta.
- **e.** anterior pituitary.

9. Decidual cells are found in the—
   a. uterus.
   b. prostate.
   c. testes.
   d. placenta.
   e. anterior pituitary.

10. ACTH does *not* stimulate the secretion of 17β-estradiol from the ovary, because—
    a. the estradiol-secreting cells lack cyclic AMP.
    b. the estradiol-secreting cells have no endoplasmic reticulum.
    c. the adenylate cyclase in the estradiol-secreting cells differs from that in the cells of the adrenal cortex.
    d. ACTH has a selective stimulatory effect on 21β-hydroxylation.
    e. the ovarian cells that secrete 17β-estradiol lack ACTH receptors.

11. Which of the labeled structures in Fig 23–3 has a higher concentration of angiotensin II than plasma?

   a   b   c   d   e

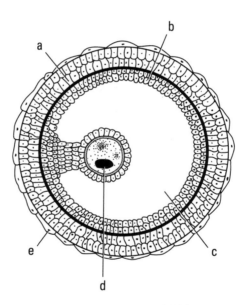

Figure 23–3. Ovarian follicle.

12. Which of the labeled structures in Fig 23–3 produces androstenedione?

   a   b   c   d   e

13. Which of the labeled structures in Fig 23–3 produces estradiol that enters the follicular fluid?

   a   b   c   d   e

14. Which of the labeled structures in Fig 23–3 contains the smallest amount of DNA per cell?

   a   b   c   d   e

In the following questions, one or more than one of the answers may be correct.
Select—

> **A** if (1), (2), and (3) are correct;
> **B** if (1) and (3) are correct;
> **C** if (2) and (4) are correct;
> **D** if only (4) is correct; and
> **E** if all are correct.

15. Which of the following are steroids?
 (1) 17α-hydroxyprogesterone.
 (2) estrone.
 (3) pregnenolone.
 (4) relaxin.

          A   **B**   **C**   **D**   **E**

16. The placenta secretes—
 (1) estrogens.
 (2) renin.
 (3) progesterone.
 (4) relaxin.

          A   **B**   **C**   **D**   **E**

17. Which of the following contain an ovum?
 (1) an immature follicle.
 (2) an atretic follicle.
 (3) a mature follicle.
 (4) a corpus luteum.

          A   **B**   **C**   **D**   **E**

18. At or immediately before the time of ovulation in humans, there is a marked increase in the plasma concentration of—
 (1) FSH.
 (2) LH.
 (3) 17β-estradiol.
 (4) progesterone.

          A   **B**   **C**   **D**   **E**

19. Which of the following are male secondary sexual characteristics?
 (1) a deep voice.
 (2) a beard.
 (3) an enlarged penis.
 (4) an increased incidence of acne.

          A   **B**   **C**   **D**   **E**

20. The fetal testis—
 (1) secretes relaxin.
 (2) secretes a factor that represses development of female internal genitalia.

(3) is under positive feedback control by pituitary LH and FSH.
(4) secretes testosterone.

A   B   C   D   E

21. Secondary amenorrhea can be caused by—
    (1) competitive athletics.
    (2) stress.
    (3) hyperprolactinemia.
    (4) hypothalamic disease.

A   B   C   D   E

22. In males, LH—
    (1) acts on receptors in the cytosol of its target cells.
    (2) does not fluctuate as much as it does in females.
    (3) is secreted at a constant, steady rate because LHRH is secreted by the hypothalamus at a constant, steady rate.
    (4) is generally under negative feedback control by gonadal steroids.

A   B   C   D   E

23. The increase in estradiol and progesterone secretion during the luteal phase of the menstrual cycle—
    (1) causes ovulation.
    (2) is associated with a rise in body temperature.
    (3) is due to secretion of these hormones by the maturing follicle.
    (4) is accentuated if pregnancy supervenes.

A   B   C   D   E

24. The hormones responsible for normal development of the mammary glands and lactation include—
    (1) cortisol.
    (2) estrogen.
    (3) insulin.
    (4) growth hormone.

A   B   C   D   E

25. Normal sexual development in the human male fetus depends on—
    (1) the presence of müllerian inhibiting substance (MIS).
    (2) secretion of testosterone.
    (3) testosterone receptors.
    (4) the presence of 5α-reductase.

A   B   C   D   E

26. In the menstrual cycle, estrogen—
    (1) decreases uterine contractility.
    (2) stimulates secretion of thick cervical mucus.

(3) raises the body temperature.
(4) stimulates growth of the endometrium.

A   B   C   D   E

27. Progesterone—
    (1) inhibits LH secretion.
    (2) stimulates secretion by the glands in the endometrium.
    (3) increases growth of lobules in the breasts.
    (4) decreases uterine responsiveness to oxytocin.

A   B   C   D   E

28. Androgens are secreted by the—
    (1) ovary.
    (2) testis.
    (3) adrenal cortex.
    (4) adrenal medulla.

A   B   C   D   E

In the following questions, indicate whether the item on the left is greater than **(G)**, the same as **(S)**, or less than **(L)** the item on the right.

29. Size of the clitoris in a woman with congenital 17α-hydroxylase deficiency.   **G   S   L**   Size of the clitoris in a woman with congenital 21β-hydroxylase deficiency.

30. Size of the clitoris in a woman with congenital 11β-hydroxylase deficiency.   **G   S   L**   Size of the clitoris in a woman with congenital 21β-hydroxylase deficiency.

31. Plasma FSH concentration in the proliferative phase of the menstrual cycle.   **G   S   L**   Plasma FSH concentration in the luteal phase of the menstrual cycle.

32. Amount of dopamine in the portal hypophyseal blood before injection of prolactin.   **G   S   L**   Amount of dopamine in the portal hypophyseal blood after injection of prolactin.

33. Adult height of a boy castrated before puberty.   **G   S   L**   Adult height of a boy with precocious puberty.

34. Plasma LH concentration in a 59-year-old man.   **G   S   L**   Plasma LH concentration in a 59-year-old woman.

35. Plasma prolactin concentration in a mother after suckling her infant.   **G   S   L**   Plasma prolactin concentration in a mother before suckling her infant.

36. Sperm production when the testicular temperature is maintained at 32 °C.   **G   S   L**   Sperm production when the testicular temperature is maintained at 37 °C.

37. hCG production in the first trimester of pregnancy.   **G   S   L**   hCG production in the third trimester of pregnancy.

38. hCS production in the first trimester of pregnancy.   **G   S   L**   hCS production in the third trimester of pregnancy.

REFERENCES

Findlay ALR: *Reproduction and the Fetus*. Arnold, 1984.

Other Endocrine Organs

24

Chapter 24 is concerned with renin and erythropoietin, 2 of the 3 hormones produced by the kidneys; atrial natriuretic peptide, the hormone produced by the heart; and melatonin, the pineal "hormone." Other putative natriuretic hormones are also considered. The formation of angiotensin II by the sequential action of renin and angiotensin-converting enzyme, the actions of angiotensin II, and the regulation of renin secretion are discussed. In addition, the effects of renal and extrarenal erythropoietin are considered. The chemistry and actions of atrial natriuretic peptide and the probable mechanisms that regulate its secretion are outlined, and the possible functions of melatonin are analyzed.

OBJECTIVES

The material in the chapter should help students to—

- Outline the reactions that lead to the formation of angiotensin II in the circulation, and describe the properties of the enzymes and substrates involved.

- Name the enzymes involved in the metabolism of angiotensin II and the products that are formed.

- List the functions of angiotensin II and the sites at which it acts to carry out these functions.

- Describe the juxtaglomerular apparatus, and list the factors that regulate its secretion.

- List drugs that affect the renin-angiotensin system and their mechanisms of action.

- Name the sources of erythropoietin in the body.

- Describe the site and mechanism of action of erythropoietin.

- Describe the structure and function of atrial natriuretic peptide.

- Diagram the steps involved in the formation of melatonin from serotonin in the pineal gland, list the proposed functions of melatonin, and discuss the regulation of melatonin secretion.

GENERAL QUESTIONS

1. Discuss the renin-angiotensin system from the point of view of the similarities and differences between it and the kallikrein system.

2. What is the major stimulus to erythropoietin secretion? How does it trigger the increase in the secretion of this hormone? Why are patients with chronic renal disease anemic?

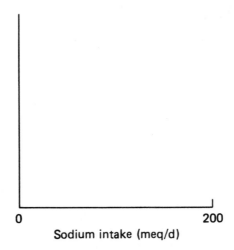

Figure 24–1.

3. Using the axes in Fig 24–1, plot the relationship between dietary sodium intake and (1) plasma renin activity, (2) the plasma concentration of atrial natriuretic peptide, and (3) the plasma concentration of cortisol.

4. How is angiotensin III formed, and what is its physiologic significance?

5. Discuss the relationship of the renin-angiotensin system to clinical hypertension.

6. List the organs in the body which appear to have renin-angiotensin systems separate from the renin-angiotensin system that produces circulating angiotensin II. What are the possible functions of the systems in these organs?

7. Define Goldblatt hypertension, Bartter's syndrome, and secondary hyperaldosteronism.

8. Discuss natriuretic hormones from the point of view of the types that may exist, their structure, their origins, and their effects.

9. One of the 2 curves in Fig 24–2 was obtained without and one was obtained with saralassin, 10^{-6} M, in the medium. Which curve is which? Explain the shape of the curves and the horizontal displacement of one from the other.

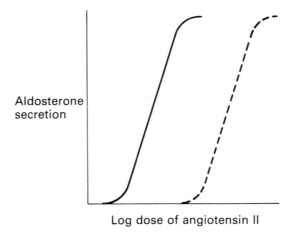

Log dose of angiotensin II

Figure 24–2. Aldosterone secretion by adrenocortical tissue in vitro.

10. Melatonin secretion is high at night and low during the day. How does the pineal gland ''know'' that it is day or night?

MULTIPLE-CHOICE QUESTIONS

In the following questions, select the single best answer.

1. Renin is secreted by—
   a. cells in the macula densa.
   b. cells in the proximal tubules.
   c. cells in the renal glomeruli.
   d. juxtaglomerular cells.
   e. renal medullary cells.

2. Erythropoietin is secreted by—
   a. cells in the macula densa.
   b. cells in the proximal tubules.
   c. cells in the renal glomeruli.
   d. juxtaglomerular cells.
   e. renal medullary cells.

3. Melatonin secretion would probably *not* be increased by—
   a. stimulation of the superior cervical ganglia.
   b. intravenous infusion of tryptophan.
   c. intravenous infusion of epinephrine.
   d. stimulation of the optic nerve.
   e. induction of pineal hydroxyindole-O-methyltransferase.

4. When a woman who has been on a low-sodium diet for 8 days is given an intravenous injection of captopril, a drug that inhibits angiotensin converting enzyme, one would expect—
   a. her blood pressure to rise because her cardiac output would fall.
   b. her blood pressure to rise because her peripheral resistance would fall.
   c. her blood pressure to fall because her cardiac output would fall.
   d. her blood pressure to fall because her peripheral resistance would fall.
   e. her plasma renin activity to fall because her circulating angiotensin I level would rise.

5. Which of the following would be expected to cause an increase in angiotensinogen secretion?
   a. vasopressin.
   b. prolactin.
   c. growth hormone.
   d. β-endorphin.
   e. ACTH.

In the following questions, one or more than one of the answers may be correct. Select—

> **A** if (1), (2), and (3) are correct;
> **B** if (1) and (3) are correct;
> **C** if (2) and (4) are correct;
> **D** if only (4) is correct; and
> **E** if all are correct.

6. Renin secretion is decreased by—
   (1) administration of a drug that blocks angiotensin converting enzyme.

(2) administration of a drug that blocks β-adrenergic receptors.
(3) narrowing the aorta between the celiac artery and the renal arteries.
(4) administration of an oral contraceptive drug containing estrogen and progesterone.

A   B   C   D   E

7. Angiotensin II increases blood pressure by an action on—
   (1) aldosterone secretion.
   (2) vascular smooth muscle.
   (3) the sympathetic nervous system.
   (4) the appetite for salt.

A   B   C   D   E

8. Administration of a drug that inhibits the action of renin would be expected to—
   (1) increase the plasma aldosterone level.
   (2) decrease the heart rate.
   (3) increase the plasma bradykinin level.
   (4) decrease the plasma angiotensin I level.

A   B   C   D   E

9. Erythropoietin—
   (1) contains zinc.
   (2) is an important ligand for iron in the tissues.
   (3) stimulates renin secretion.
   (4) acts on some but not all stem cells in bone marrow.

A   B   C   D   E

10. Which of the following clinical and experimental conditions increases the secretion of atrial natriuretic peptide?
    (1) primary hyperaldosteronism.
    (2) infusion of isotonic saline.
    (3) infusion of hyperoncotic albumin.
    (4) constriction of the ascending aorta close to the heart.

A   B   C   D   E

In the following questions, indicate whether the item on the left is greater than **(G)**, the same as **(S)**, or less than **(L)** the item on the right.

11. Increase in melatonin secretion produced by norepinephrine.   **G S L**   Increase in melatonin secretion produced by acetylcholine.

12. Sodium excretion after injection of an extract of cardiac atria.   **G S L**   Sodium excretion after injection of an extract of cardiac ventricles.

13. Plasma renin *activity* before injection of estrogen.   **G S L**   Plasma renin *activity* after injection of estrogen.

14. Plasma renin *concentration* before injection of estrogen.

G S L

Plasma renin *concentration* after injection of estrogen.

15. Rate of erythropoietin secretion at an altitude of 1000 meters.

G S L

Rate of erythropoietin secretion at an altitude of 10,000 meters.

16. Amount of erythropoietin in the blood of a patient with chronic respiratory failure.

G S L

Amount of erythropoietin in the blood of a patient with chronic renal failure.

REFERENCES

Cantin M, Genest J: The heart and the atrial natriuretic factor. *Endocr Rev* 1985;**6**:107.

Keeton K, Campbell WB: The pharmacologic alteration of renin secretion. *Pharmacol Rev* 1980;**32**:81.

Digestion & Absorption

The physiology of the gastrointestinal system is considered in Chapters 25 and 26. Chapter 25 is concerned with digestion and absorption, and Chapter 26 deals with the details of how the gastrointestinal tract and its associated glands function to aid digestion and absorption. The products formed from ingested food in the gastrointestinal tract are analyzed in Chapter 25, along with the absorption of the products of digestion plus cholesterol, water, electrolytes, vitamins, calcium, and iron. In addition to the questions for this chapter, integrated questions concerned with material covered in both chapters are included in the questions for Chapter 26.

OBJECTIVES

The material in the chapter should help students to—

- Understand the overall functions of the gastrointestinal system.

- List the principal digestive enzymes, their precursors, their substrates, and the products of the action of the enzymes.

- Define brush border, unstirred layer, and glycocalyx.

- Describe the conversion of dietary carbohydrates into glucose and other hexoses ready for absorption from the intestine.

- Summarize the processes involved in the absorption of hexoses and pentoses from the intestine into the bloodstream.

- Describe the conversion of dietary protein into amino acids and small peptides ready for absorption from the intestine.

- Summarize the processes involved in the absorption of amino acids and small peptides from the intestine into the bloodstream.

- Outline the events occurring during digestion of fats.

- Describe the processes by which fatty acids and other lipids are absorbed from the intestine into the bloodstream.

- Summarize the processes that regulate the absorption of water, Na^+, K^+, and HCO_3^- from the gastrointestinal tract.

- Outline the processes that regulate calcium absorption.

- Outline the processes that regulate iron absorption.

GENERAL QUESTIONS

1. Why do patients with gastrin-secreting tumors of the pancreas (Zollinger-Ellison syndrome) sometimes develop steatorrhea?

2. Many adults who are not of western European origin have abdominal symptoms after ingestion of milk. What are the symptoms, what causes them, and how would you treat the condition?

3. Describe secondary active transport, the process producing movement of glucose from the intestinal lumen to the bloodstream. Why is it called secondary rather than primary active transport?

4. After a meal, the lumens of the stomach and the intestine contain powerful proteolytic enzymes. Why do these enzymes not digest the pancreas, which produces them, or the wall of the intestine?

5. Compare and contrast the absorption of glucose, amino acids, and fatty acids.

6. Intestinal absorption of calcium and iron are both subject to feedback control but by very different mechanisms. Compare and contrast the mechanisms.

7. What are M cells? What is their function?

MULTIPLE-CHOICE QUESTIONS

In the following questions, select the single best answer.

1. The pathway from the intestinal lumen to the circulating blood for a short-chain fatty acid (< 10 carbon atoms) is—
    a. intestinal mucosal cell→chylomicrons→lymphatic duct→systemic venous blood.
    b. intestinal mucosal cell→hepatic portal vein blood→systemic venous blood.
    c. space between mucosal cells→lymphatic duct→systemic venous blood.
    d. space between mucosal cells→chylomicrons→lymphatic duct→systemic venous blood.
    e. intestinal mucosal cell→LDL→hepatic portal vein blood→systemic venous blood.

2. The exocrine portion of the pancreas does *not* secrete—
    a. chymotrypsinogen.
    b. proelastase.
    c. α-limit dextrinase.
    d. pancreatic α-amylase.
    e. deoxyribonuclease.

3. Water is absorbed in the jejunum, ileum, and colon and excreted in the feces. Arrange these in order of the amount of water absorbed or excreted from greatest to smallest.
    a. colon, jejunum, ileum, feces.
    b. feces, colon, ileum, jejunum.
    c. jejunum, ileum, colon, feces.
    d. colon, ileum, jejunum, feces.
    e. feces, jejunum, ileum, colon.

4. Drugs and toxins that increase the cyclic AMP content of the intestinal mucosa cause diarrhea because they—
    a. increase Na^+-H^+ cotransport in the small intestine.
    b. increase K^+ secretion in the colon.
    c. inhibit potassium absorption in the crypts of Lieberkühn.

    d.   increase absorption of Na^+ in the small intestine.
    e.   increase secretion of Cl^- into the intestinal lumen.

  5. The brush border of intestinal cells contains large amounts of—
    a.   pepsin.
    b.   maltase.
    c.   secretin.
    d.   salivary α-amylase.
    e.   carboxypeptidase B.

In the following questions, one or more than one of the answers may be correct.
Select—

        A if (1), (2), and (3) are correct;
        B if (1) and (3) are correct;
        C if (2) and (4) are correct;
        D if only (4) is correct; and
        E if all are correct.

  6. Which of the following are *correctly* paired?
    (1) pancreatic α-amylase : starch.
    (2) elastase : tissues rich in elastin.
    (3) enteropeptidase : trypsinogen.
    (4) rennin : coagulated milk.

            A   **B**   **C**   **D**   **E**

  7. The ileum, not the jejunum, normally absorbs—
    (1) iron.
    (2) cyanocobalamin (vitamin B_{12}).
    (3) fructose.
    (4) bile salts.

            A   **B**   **C**   **D**   **E**

  8. The following processes are involved in the movement of the products of
digestion from the lumen of the gastrointestinal tract to the blood—
    (1) facilitated diffusion.
    (2) secondary active transport.
    (3) diffusion.
    (4) endocytosis.

            A   **B**   **C**   **D**   **E**

  9. Commonly ingested glucose polymers include—
    (1) amylopectin.
    (2) maltotriose.
    (3) glycogen.
    (4) maltose.

            A   **B**   **C**   **D**   **E**

 10. Calcium absorption is increased by—
    (1) 1,25-dihydroxycholecalciferol.
    (2) oxalates in the diet.
    (3) hypocalcemia.
    (4) hypercalcemia.

            A   **B**   **C**   **D**   **E**

In the following questions, indicate whether the item on the left is greater than **(G)**, the same as **(S)**, or less than **(L)** the item on the right.

11. Na⁺ absorption when there is glucose in the intestinal lumen.

    G S L

    Na⁺ absorption when there is no glucose in the intestinal lumen.

12. Na⁺ absorption when there are amino acids in the intestinal lumen.

    G S L

    Na⁺ absorption when there are no amino acids in the intestinal lumen.

13. Amount of glucose absorbed from the intestine following intravenous administration of insulin.

    G S L

    Amount of glucose absorbed from the intestine following intravenous administration of glucagon.

14. Absorption of amino acids in the jejunum.

    G S L

    Absorption of amino acids in the ascending colon.

15. Absorption of whole proteins from the gastrointestinal tract in infants.

    G S L

    Absorption of whole proteins from the gastrointestinal tract in adults.

16. Absorption of fatty acids when micelles are present in the intestinal lumen.

    G S L

    Absorption of fatty acids when micelles are absent from the intestinal lumen.

17. Absorption of fatty acids in the stomach.

    G S L

    Absorption of fatty acids in the small intestine.

18. Amount of water reabsorbed from the gastrointestinal tract per day.

    G S L

    Amount of water ingested per day.

19. Absorption of vitamin A when bile duct is obstructed.

    G S L

    Absorption of vitamin A when bile duct is patent.

20. Iron absorption in anemia.

    G S L

    Iron absorption in hemochromatosis.

REFERENCES

Davenport HW: *A Digest of Digestion*, 2nd ed. Year Book, 1978.

26

Regulation of Gastrointestinal Function

The subject of Chapter 26 is a detailed consideration of the way the gastrointestinal tract and its associated glands, the salivary glands, liver, and pancreas function in carrying out digestion and absorption of food. The gastrointestinal hormones are first reviewed, and then the functions of the mouth and salivary glands are analyzed. Regulation of gastric secretion and motility are considered, along with regulation of pancreatic exocrine secretion. The multiple functions of the liver are summarized, and bile secretion and the functions of the gallbladder are reviewed. The functional anatomy of the small intestine is analyzed, and regulation of intestinal motility and secretion is discussed, along with a brief consideration of the malabsorption syndrome and intestinal obstruction. Regulation of the motility of the colon and defecation are discussed, and the effects of the bacteria in the colon and a number of colonic diseases are considered.

OBJECTIVES

The material in the chapter should help students to—

- Summarize the structure and innervation of the gastrointestinal tract.

- List the principal gastrointestinal hormones, and, for each hormone, list the sites where it is secreted and its main physiologic function.

- Summarize the functions of the mouth, the salivary glands, and the esophagus.

- Describe the stomach in terms of its functional anatomy and histology.

- Describe how acid is secreted by cells in the gastric mucosa.

- Describe the mechanisms that regulate the secretion and motility of the stomach.

- Describe the relationship between cyanocobalamin and intrinsic factor.

- List the main components of pancreatic juice, and outline the mechanisms that regulate its secretion.

- Describe the functional anatomy of the liver, and discuss the formation of bile.

- Discuss the function of the gallbladder and the processes that regulate the passage of bile to the intestinal lumen.

- List the types of movement seen in the small intestine and the function of each.

- Name the products secreted by the intestinal mucosa.

- List the types of movement seen in the colon and the function of each.

- Describe the events in the colon that lead to defecation, and outline the neural pathways that control this semireflex response.

GENERAL QUESTIONS

1. Compare and contrast the mechanisms by which short-chain and long-chain fatty acids are absorbed from the intestine.

2. What factors regulate the secretion of the exocrine portion of the pancreas?

3. Discuss the abnormalities of gastrin secretion seen in disease states.

4. What abnormalities would you expect to be produced by resection of the terminal portion of the ileum? Why?

5. Compare the small and the large intestine in terms of structure and function.

6. What is achalasia? Discuss its pathophysiology, and suggest ways it might be treated.

7. What causes gallstones? How would you treat them?

8. Several surgical procedures have been recommended for the treatment of severe obesity that fails to respond to other forms of treatment. What are these procedures? How do they cause weight loss? What are their long-term complications?

9. The bacteria in the colon and their host exist in a symbiotic relationship. How does the host benefit from this relationship, and what harmful or potentially harmful effects may occur to the host?

10. Discuss the pathophysiology of constipation and diarrhea.

MULTIPLE-CHOICE QUESTIONS

In the following questions, select the single best answer.

1. Which of the labeled structures in Fig 26–1 is the site where maltase is found?

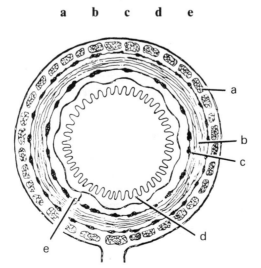

Figure 26–1. Cross section of small intestine. (Reproduced with permission from Bell GH, Emslie-Smith D, Paterson CR: *Textbook of Physiology and Biochemistry*, 9th ed. Churchill Livingstone, 1976.)

2. Which of the labeled structures in Fig 26–1 is the site where vasoactive intestinal polypeptide (VIP) is found?

a   b   c   d   e

3. Gastrin secretion is increased by—
   a. acid in the lumen of the stomach.
   b. distention of the stomach.
   c. increased circulating levels of secretin.
   d. increased circulating levels of GIP.
   e. vagotomy.

4. Gastrin is structurally similar to—
   a. GIP.
   b. VIP.
   c. glucagon.
   d. cholecystokinin (CCK).
   e. secretin.

5. In infants, defecation often follows a meal. The cause of colonic contractions in this situation is—
   a. the gastroileal reflex.
   b. increased circulating levels of CCK.
   c. the gastrocolic reflex.
   d. increased circulating levels of somatostatin.
   e. the enterogastric reflex.

6. Glucose absorption is increased to the greatest degree by—
   a. injection of insulin.
   b. drinking isotonic saline.
   c. drinking a solution containing fructose.
   d. injection of glucagon.
   e. drugs that inhibit Na^+, K^+ ATPase.

7. After a meal rich in carbohydrate is ingested, insulin secretion is probably stimulated by—
   a. GIP.
   b. CCK.
   c. serotonin.
   d. VIP.
   e. gastrin.

8. The symptoms of the dumping syndrome (discomfort after meals in patients with intestinal short circuits such as anastamosis of the jejunum to the stomach) are caused in part by—
   a. increased blood pressure.
   b. increased secretion of glucagon.
   c. increased secretion of CCK.
   d. hypoglycemia.
   e. hyperglycemia.

9. Which of the following has the highest pH?
   a. gastric juice.
   b. bile in the gallbladder.
   c. pancreatic juice.
   d. saliva.
   e. secretions of the intestinal glands.

10. Which of the following is *least* likely to be found in saliva?

    a.   salivary α-amylase.
    b. Na^+.
    c.   renin.
    d. HCO_3^-.
    e.   chymotrypsinogen.

The next 4 questions refer to Table 26–1.

Table 26–1.

Pattern	Plasma Bilirubin		Plasma Alkaline Phosphatase	Hematocrit	Plasma Bile Acids
	Direct	Indirect			
a.	—	—	—	—	↓
b.	↑	↑↑	—	—	—
c.	↑↑	↑	↑	—	↓
d.	↑	↑↑	—	↓	—
e.	↑	↑↑	↑	—	↑

— = no change; ↑ = increase; ↑↑ = marked increase; ↓ = decrease.

11.   Which of the sets of data in Table 26–1 would be found in a patient with a gallstone obstructing the common bile duct?

              a   b   c   d   e

12.   Which set of data would be found in a patient with intravascular hemolysis?

              a   b   c   d   e

13.   Which set of data would be found in a patient with infectious hepatitis?

              a   b   c   d   e

14.   Which set of data would be found in a patient with a resection of the ileum?

              a   b   c   d   e

In the following questions, one or more than one of the answers may be correct. Select—

              A if (1), (2), and (3) are correct;
              B if (1) and (3) are correct;
              C if (2) and (4) are correct;
              D if only (4) is correct; and
              E if all are correct.

15.   Which of the following functions would be disrupted by a gallstone blocking the common bile duct?
    (1) formation of the micelles involved in fat absorption in the intestine.
    (2) emulsification of fats in the intestine.
    (3) removal of cholesterol from the circulation.
    (4) absorption of carbohydrates.

             A   **B**   **C**   **D**   **E**

16. Which of the following hormones are found in the mucosa of the gastrointestinal tract and in the brain?
    (1) CCK.
    (2) glucagon.
    (3) VIP.
    (4) ACTH.

    A   B   C   D   E

17. Peristaltic waves—
    (1) do not occur in the absence of the intrinsic innervation of the gastrointestinal tract.
    (2) can be initiated by stretching a portion of the intestinal wall.
    (3) are unaffected by morphine.
    (4) can be influenced by the sympathetic nervous system.

    A   B   C   D   E

18. Which of the following effects is produced by a hormone secreted by the mucosa of the gastrointestinal tract?
    (1) increased production of pancreatic juice rich in enzymes.
    (2) increased secretion of glucagon.
    (3) increased production of watery pancreatic juice rich in bicarbonate.
    (4) decreased gastric motility.

    A   B   C   D   E

19. Which of the following are important in mixing the contents of the gastrointestinal tract?
    (1) peristalsis in the esophagus.
    (2) segmentation contractions.
    (3) mass action contractions.
    (4) peristalsis in the stomach.

    A   B   C   D   E

20. The lower esophageal sphincter—
    (1) normally prevents reflux of gastric contents into the esophagus.
    (2) relaxes ahead of a peristaltic wave passing down the esophagus.
    (3) is abnormally contracted in achalasia.
    (4) is in spasm in patients with heartburn.

    A   B   C   D   E

21. Gastric secretion is increased by—
    (1) anger and hostility.
    (2) acid in the duodenum.
    (3) secretin.
    (4) stimulation of the vagus nerves.

    A   B   C   D   E

22. Removal of the entire colon would be expected to cause—
    (1) death.
    (2) severe malnutrition.
    (3) megaloblastic anemia.
    (4) a decrease in the blood level of ammonia in patients with cirrhosis of the liver.

    A   B   C   D   E

23. The hormones that normally regulate the secretion of the exocrine portion of the pancreas include—
    (1) gastrin.
    (2) CCK.
    (3) GIP.
    (4) secretin.

    A   B   C   D   E

24. Which of the following would be expected to decrease gastric emptying?
    (1) the enterogastric reflex.
    (2) gastrin.
    (3) GIP.
    (4) distention of the stomach.

    A   B   C   D   E

25. Bile contains—
    (1) bile acids.
    (2) cholesterol.
    (3) bilirubin glucuronide.
    (4) alkaline phosphatase.

    A   B   C   D   E

26. Megaloblastic anemia might be expected to occur following removal of—
    (1) the stomach.
    (2) the duodenum.
    (3) the terminal ileum.
    (4) the ascending colon.

    A   B   C   D   E

27. Total pancreatectomy would be expected to cause—
    (1) decreased absorption of fat-soluble vitamins.
    (2) little or no change in the plasma glucagon level.
    (3) a rise in the blood glucose level.
    (4) decreased absorption of amino acids.

    A   B   C   D   E

28. After complete hepatectomy, a rise would be expected in the blood level of—
    (1) fibrinogen.
    (2) 25-hydroxycholecalciferol.

(3) glucose.
(4) ammonia.

A   B   C   D   E

In the following questions, indicate whether the item on the left is greater than
(G), the same as **(S)**, or less than **(L)** the item on the right.

29.	Concentration of glucose in hepatic portal blood after a meal.	G S L	Concentration of glucose in systemic blood after a meal.
30.	Concentration of amino acids in hepatic portal blood after a meal.	G S L	Concentration of amino acids in systemic blood after a meal.
31.	Concentration of albumin-bound long-chain fatty acids in hepatic portal blood after a meal.	G S L	Concentration of albumin-bound long-chain fatty acids in systemic blood after a meal.
32.	Gastric acid secretion with fat in the duodenum.	G S L	Gastric acid secretion with polypeptides and amino acids in the duodenum.
33.	Intestinal motility following stimulation of sympathetic nerves to the intestine.	G S L	Intestinal motility following stimulation of parasympathetic nerves to the intestine.
34.	Esophageal peristalsis after swallowing in an individual who is standing upright.	G S L	Esophageal peristalsis after swallowing in an individual who is standing on his or her head.
35.	Blood glucose level after total cholecystectomy.	G S L	Blood glucose level after total gastrectomy.
36.	Fat absorption after taking a drug that inhibits protein synthesis.	G S L	Fat absorption after taking a drug that stimulates protein synthesis.

REFERENCES

Davenport HW: *Physiology of the Digestive Tract*, 5th ed. Year Book, 1982.

Circulating Body Fluids

27

In Chapter 27, the functions of blood and lymph are reviewed. The formed elements of the blood—white blood cells, red blood cells, and platelets—are described, and the physiologic role of each is considered. This includes a review of immune mechanisms and their role in the rejection of foreign substances. The composition of plasma and lymph is also discussed, along with an analysis of blood types and the clotting and anticlotting mechanisms.

OBJECTIVES

The material in the chapter should help students to—

- Describe in general terms the systemic, pulmonary, and lymphatic divisions of the circulatory system.

- Name the types of bone marrow, their function, and where they are located.

- List the various types of cells found in the blood and the precursor cells from which each develops.

- Describe the functions of neutrophils.

- Describe the functions of monocytes.

- Name the various types of lymphocytes, and describe the origin and function of each.

- Describe the processes responsible for the production of circulating antibodies and for the rejection of tissue transplanted from other individuals.

- Describe the structure and function of platelets and the way they discharge their granules.

- Outline the functions of red blood cells.

- Name the common blood types, and describe how blood is typed and cross-matched.

- Name the principal substances found in blood plasma.

- Describe the events that lead to the clotting of blood by the intrinsic and extrinsic pathways.

GENERAL QUESTIONS

1. Compare and contrast the composition of blood plasma and lymph.

2. How is the production of white blood cells, platelets, and red blood cells adjusted to meet the varying needs of the individual?

3. What are the tissue macrophages, and what is their origin and function?

4. Mutant genes that cause the production of abnormal hemoglobins are common in humans. What determines whether a given mutation in hemoglobin is harmless or harmful?

5. Describe how blood and tissue typing can be used to determine paternity.

6. What is the fibrinolytic system, and what is its function?

7. What is the clonal selection theory of antibody formation? What is the evidence for this theory?

8. Why is blood clotting abnormal in patients with vitamin K deficiency?

9. What are the functions of natural killer cells? How do these cells differ from other lymphocytes?

10. Compare and contrast humoral and cellular immunity.

MULTIPLE-CHOICE QUESTIONS

In the following questions, select the single best answer.

1. Which of the patterns in Fig 27–1 would be seen if plasma from an individual with blood type B were mixed with red cells from an individual with blood type O?

a   b

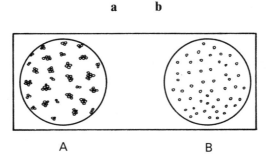

A   B

Figure 27–1. Red blood cells in plasma.

2. Which of the patterns seen in Fig 27–1 would be seen if plasma from an individual with blood type O were added to red cells from an individual with blood type B?

a   b

3. In Fig 27–1, the plasma hemoglobin would be highest in which pattern?

a   b

4. Which of the following is *not* primarily a function of blood plasma?
   a. transport of hormones.
   b. maintenance of red cell size.
   c. transport of chylomicrons.
   d. transport of antibodies.
   e. transport of O_2.

5. A hematocrit of 41% means that, in the sample of blood analyzed—
   a. 41% of the hemoglobin is in the plasma.
   b. 41% of the total blood volume is made up of blood plasma.
   c. 41% of the total blood volume is made up of red and white blood cells.
   d. 41% of the hemoglobin is in red blood cells.
   e. 41% of the formed elements in blood are red blood cells.

6. In normal human blood—
   a. the eosinophil is the most common type of white blood cell.
   b. there are more lymphocytes than neutrophils.
   c. the iron is mostly in hemoglobin.
   d. there are more white cells than red cells.
   e. there are more lymphocytes than red cells.

7. Lymphocytes—
   a. all originate from the bone marrow.
   b. are unaffected by hormones.
   c. convert to monocytes in response to antigens.
   d. interact with eosinophils to produce platelets.
   e. are part of the body's defense against cancer.

In the following questions, one or more than one of the answers may be correct. Select—

A if (1), (2), and (3) are correct;
B if (1) and (3) are correct;
C if (2) and (4) are correct;
D if only (4) is correct; and
E if all are correct.

8. Production of red blood cells is increased by—
   (1) high oxygen tension in the blood.
   (2) low oxygen tension in the blood.
   (3) angiotensinogen.
   (4) erythropoietin.

A B C D E

9. Platelets—
   (1) are formed from megakaryocytes.
   (2) have a high serotonin content.
   (3) are affected by derivatives of arachidonic acid.
   (4) have large nuclei.

A B C D E

10. Cells of the reticuloendothelial system are found in the—
    (1) lungs.
    (2) liver.
    (3) bone marrow.
    (4) stomach.

A B C D E

11. B lymphocytes—
    (1) are the precursors of plasma cells.
    (2) are the precursors of killer lymphocytes.

(3) form clones of antibody-generating cells.
(4) produce lymphokines.

A   B   C   D   E

12. Which of the following diseases are due to abnormalities of the normal clotting mechanisms?
(1) thrombocytopenic purpura.
(2) hemophilia A.
(3) Christmas disease.
(4) sickle cell anemia.

A   B   C   D   E

13. The cell shown in Fig 27–2—
(1) contains actin.
(2) is capable of passing through the walls of blood vessels.
(3) contains acid proteases.
(4) is unaffected by glucocorticoid hormones.

A   B   C   D   E

14. Production of the type of cell shown in Fig 27–2 is increased by—
(1) interleukin-3 (IL-3).
(2) macrophage colony-stimulating factor (M-CSF).
(3) granulocyte-macrophage colony-stimulating factor (GM-CSF).
(4) erythropoietin.

A   B   C   D   E

Figure 27–2.

In the following questions, match each numbered item (immunoglobulin) with the lettered items (characteristics) below that are most closely associated with it. Lettered items may be selected once, more than once, or not at all.
(A) Complement fixation.
(B) Secretory immunity.

(C) Contains J chain.
(D) Contains SC chain.
(E) Monomer only.

15. IgA   **A   B   C   D   E**

16. IgD   **A   B   C   D   E**

17. IgE   **A   B   C   D   E**

18. IgG   **A   B   C   D   E**

19. IgM   **A   B   C   D   E**

In the following questions, indicate whether the item on the left is greater than **(G)**, the same as **(S)**, or less than **(L)** the item on the right.

20. Blood hemoglobin concentration in anemia.   **G   S   L**   Blood hemoglobin concentration in polycythemia.

21. Number of light chains in a basic immunoglobulin molecule.   **G   S   L**   Number of heavy chains in a basic immunoglobulin molecule.

22. Tendency to hemorrhage when the platelet count is high.   **G   S   L**   Tendency to hemorrhage when the platelet count is low.

23. Percent of total plasma osmolality due to circulating albumin.   **G   S   L**   Percent of total plasma osmolality due to circulating sodium chloride.

24. Incidence of the sickle cell gene in North America.   **G   S   L**   Incidence of the sickle cell gene in Africa.

25. Total O_2 per unit of blood in the aorta.   **G   S   L**   Total O_2 per unit of blood in the inferior vena cava.

REFERENCES

MacKinney AA (editor): *Pathophysiology of Blood.* Wiley, 1984.
Stites DP, Stobo JD, Wells JV (editors): *Basic & Clinical Immunology*, 6th ed. Appleton & Lange, 1987.

28

Origin of the Heartbeat & the Electrical Activity of the Heart

Chapter 28 is a review of the conduction system of the heart and the way the impulse that generates a normal heartbeat spreads from the sinoatrial (SA) node through the atrioventricular (AV) node and from the AV node via the bundle of His and the Purkinje system to all parts of the ventricles. The electrocardiogram (ECG) and its genesis are described as well as the vectorcardiogram and the His bundle electrogram. Cardiac arrhythmias are reviewed and analyzed, and the electrocardiographic abnormalities produced by myocardial infarction and various electrolyte abnormalities are summarized.

OBJECTIVES

The material in the chapter should help students to—

- Describe the structure and function of the conduction system of the heart, and compare the action potentials in it to those in cardiac muscle.

- Describe the way the ECG is recorded, the waves of the ECG, and the relationship of the ECG to the electrical axis of the heart.

- Define His bundle electrogram, and name the waves seen in it.

- Name the common cardiac arrhythmias, and summarize the processes that produce them.

- Define reentry.

- Know how and when to carry out cardiac massage.

- List the principal early and late electrocardiographic manifestations of myocardial infarction, and explain the early changes in terms of the underlying ionic events that produce them.

- Describe the electrocardiographic changes and the changes in cardiac function produced by hyperkalemia and hypokalemia and by hypercalcemia and hypocalcemia.

GENERAL QUESTIONS

1. What is sinus arrhythmia? How is it produced? What is its clinical significance?

2. What is an ectopic focus of excitation? What is its pathophysiologic significance?

3. Discuss the clinical value of the His bundle electrogram.

4. List and explain the effects of slow channel Ca^{2+} blocking drugs on the heart. In what clinical conditions are they of value?

5. How does reentry cause abnormal cardiac rhythms?

6. Compare and contrast the Wolff-Parkinson-White syndrome and the Lown-Ganong-Levine syndrome. What are the underlying defects that produce these 2 conditions?

7. Define calcium rigor, and explain its occurrence.

MULTIPLE-CHOICE QUESTIONS

In the following questions, select the single best answer.

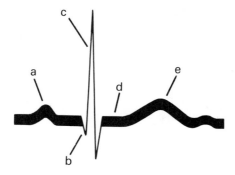

Figure 28–1. ECG.

1. Which of the labels in Fig 28–1 identifies the S-T segment?

a   b   c   d   e

2. Which of the labels in Fig 28–1 identifies the Q wave?

a   b   c   d   e

3. Which of the labels in Fig 28–1 identifies the part of the ECG that corresponds to maximum opening of Na^+ channels?

a   b   c   d   e

4. Which of the labels in Fig 28–1 identifies the part of the ECG that corresponds to maximum opening of Ca^{2+} channels?

a   b   c   d   e

5. Which of the following normally has the steepest prepotential?
  a. the sinoatrial node.
  b. the atrioventricular node.
  c. the bundle of His.
  d. the terminals of Purkinje fibers.
  e. ventricular muscle cells.

6. In second-degree heart block—
  a. the ventricular rate is slower than the atrial rate.
  b. the ventricular electrocardiographic complexes are distorted.
  c. there is a high incidence of ventricular tachycardia.
  d. stroke volume is decreased.
  e. cardiac output is increased.

7. The PR interval of the ECG corresponds to—
   a. ventricular repolarization.
   b. ventricular depolarization.
   c. repolarization of the AV node and the bundle of His.
   d. atrial depolarization and conduction through the AV node.
   e. conduction from the SA node to the muscle of the atrium.

8. Carotid sinus massage sometimes stops supraventricular tachycardia because—
   a. it decreases sympathetic discharge to the SA node.
   b. it increases vagal discharge to the SA node.
   c. it increases vagal discharge to the conducting tissue between the atria and the ventricles.
   d. it decreases sympathetic discharge to the conducting tissue between the atria and the ventricles.
   e. it increases the refractory period of the ventricular myocardium.

9. Which of the following cardiac tissues has the highest conduction velocity?
   a. the SA node.
   b. atrial muscle.
   c. the AV node.
   d. Purkinje fibers.
   e. ventricular muscle fibers.

In the following questions, one or more than one of the answers may be correct. Select—

   A if (1), (2), and (3) are correct;
   B if (1) and (3) are correct;
   C if (2) and (4) are correct;
   D if only (4) is correct; and
   E if all are correct.

10. The mean QRS vector (electrical axis of the heart) in the frontal plane—
    (1) can be determined from 3 standard limb leads.
    (2) is normally the same during systole and diastole.
    (3) is of value in determining whether left or right axis deviation is present.
    (4) may be reversed in essential hypertension.

    A   B   C   D   E

11. In complete heart block—
    (1) fainting may occur because the atria are unable to pump blood into the ventricles.
    (2) death due to ventricular fibrillation is common.
    (3) the atrial rate is slower than the ventricular rate.
    (4) fainting may occur because of prolonged periods during which the ventricles fail to contract.

    A   B   C   D   E

12. Reentry is a common cause of—
    (1) paroxysmal atrial tachycardia.
    (2) paroxysmal nodal tachycardia.
    (3) atrial fibrillation.
    (4) sinus arrhythmia.

    A   B   C   D   E

13. Which of the following electrocardiographic changes are characteristic of myocardial infarction?
    (1) the appearance of Q waves that were not previously present.
    (2) a short PR interval.
    (3) elevation of ST segments in some leads.
    (4) bradycardia.

A   **B**   **C**   **D**   **E**

14. In which of the following arrhythmias would you expect drugs that increase the refractory period of conduction tissue and muscle fibers to be of therapeutic value?
    (1) atrial fibrillation.
    (2) paroxysmal atrial tachycardia.
    (3) paroxysmal ventricular tachycardia.
    (4) ventricular fibrillation.

A   **B**   **C**   **D**   **E**

In the following questions, indicate whether the item on the left is greater than **(G)**, the same as **(S)**, or less than **(L)** the item on the right.

15. Height of T waves when the plasma K^+ level is 5.5 meq/L.   **G S L**   Height of T waves when the plasma K^+ level is 8.5 meq/L.

16. In normal adults, height of the R wave in lead V_1 of the ECG.   **G S L**   In normal adults, height of the R wave in lead V_5 of the ECG.

17. Heart rate in first-degree heart block.   **G S L**   Heart rate in second-degree heart block.

18. Interval between an atrial premature beat and the next normal beat.   **G S L**   Interval between a ventricular premature beat and the next normal beat.

19. Duration of a normal P wave.   **G S L**   Duration of a normal T wave.

REFERENCES

Goldschlager N, Goldman MJ: *Electrocardiography: Essentials of Interpretation.* Lange, 1984.

29

The Heart as a Pump

Chapter 29 describes the mechanical events in the cardiac cycle as the heart pumps the blood through the pulmonary and systemic circulations. The arterial and venous pulses are considered, and there is a review of heart sounds and murmurs. Measurement and regulation of cardiac output are also discussed.

OBJECTIVES

The material in the chapter should help students to—

- Describe the events that occur in the heart during diastole, atrial systole, isovolumetric ventricular contraction, and ventricular ejection.

- Outline the changes in the duration of systole and diastole that occur with changes in heart rate, and discuss their physiologic consequences.

- Describe the arterial pulse and jugular venous pulse.

- Describe and explain the first and second heart sounds and the occasionally observed third and fourth heart sounds.

- State the timing of the murmurs produced by aortic stenosis, aortic insufficiency, mitral stenosis, and mitral insufficiency.

- List the methods commonly used to measure cardiac output and the physiologic basis for each.

- List the factors affecting cardiac output and the effect of each.

- Summarize the factors governing oxygen consumption by the heart.

GENERAL QUESTIONS

1. The events occurring on the 2 sides of the heart are not exactly synchronous. Why does the asynchrony occur, and how is it compatible with life and health?

2. Within limits, the heart normally pumps out an amount of blood equal to the amount returned to it. What mechanisms make this possible? What happens when the heart fails to pump out the blood returning to it?

3. It takes several months for nerves to grow into transplanted hearts. However, before the nerves regrow, patients with transplanted hearts are able to increase their cardiac output when they exercise. What is the mechanism involved, and how does it operate?

4. Describe the pressure-volume loop of the ventricle, and explain how it is produced.

5. Explain postextrasystolic potentiation.

6. What is the ejection fraction? What conditions cause it to increase, and what conditons cause it to decrease? Why?

MULTIPLE-CHOICE QUESTIONS

In the following questions, select the single best answer.

1. The first heart sound is caused by—
   a. closure of the aortic and pulmonary valves.
   b. vibrations in the ventricular wall during systole.
   c. ventricular filling.
   d. closure of the mitral and tricuspid valves.
   e. retrograde flow in the vena cava.

2. The fourth heart sound is caused by—
   a. closure of the aortic and pulmonary valves.
   b. vibrations of the ventricular wall during systole.
   c. ventricular filling.
   d. closure of the mitral and tricuspid valves.
   e. retrograde flow into the vena cava.

3. In a patient with mitral stenosis, one would expect to hear—
   a. a continuous murmur.
   b. a systolic murmur loudest over the base of the heart.
   c. a systolic murmur loudest over the apex of the heart.
   d. a diastolic murmur loudest over the base of the heart.
   e. a diastolic murmur loudest over the apex of the heart.

4. The dicrotic notch on the aortic pressure curve is caused by—
   a. closure of the mitral valve.
   b. closure of the tricuspid valve.
   c. closure of the aortic valve.
   d. closure of the pulmonary valve.
   e. rapid filling of the left ventricle.

5. During exercise, a man consumes 1.8 L of oxygen per minute. His arterial O_2 content is 190 mL/L, and the O_2 content of his mixed venous blood is 134 mL/L. His cardiac output is approximately—
   a. 3.2 L/min.
   b. 16 L/min.
   c. 32 L/min.
   d. 54 L/min.
   e. 160 mL/min.

6. The work performed by the left ventricle is substantially greater than that performed by the right ventricle, because in the left ventricle—
   a. the contraction is slower.
   b. the wall is thicker.
   c. the stroke volume is greater.
   d. the preload is greater.
   e. the afterload is greater.

7. The v wave of the jugular pulse is produced by—
   a. atrial systole.
   b. the rise in ventricular pressure produced by isovolumetric ventricular contraction.
   c. the rise in atrial pressure before the tricuspid valve opens during diastole.
   d. flow from the left atrium to the right atrium through an atrial septal defect.
   e. ventricular ejection.

In the following questions, one or more than one of the answers may be correct. Select—

A if (1), (2), and (3) are correct;
B if (1) and (3) are correct;
C if (2) and (4) are correct;
D if only (4) is correct; and
E if all are correct.

8. During the cardiac cycle—
(1) myocardial contraction begins in the left atrium.
(2) myocardial contraction begins in the right atrium.
(3) excitation of the epicardial myocardium precedes excitation of the endocardial myocardium.
(4) excitation of the septum precedes excitation of the ventricular wall.

A   B   C   D   E

9. Ventricular filling—
(1) is reduced if the aortic valve is stenotic.
(2) begins during diastole.
(3) gives rise to the second heart sound.
(4) is reduced if the mitral valve is stenotic.

A   B   C   D   E

10. Stimulation of the sympathetic nerves to the heart would be expected to cause—
(1) an increase in heart rate.
(2) an increase in stroke volume.
(3) a decrease in the size of the ventricles.
(4) greater filling of the atria.

A   B   C   D   E

11. The following are primary determinants of cardiac output—
(1) stroke volume.
(2) blood flow to cardiac muscle.
(3) heart rate.
(4) hematocrit.

A   B   C   D   E

12. Starling's law of the heart—
(1) does not operate in the failing heart.
(2) is one of the factors affecting the inotropic response of the heart.
(3) is one of the factors affecting the chronotropic response of the heart.
(4) affects the end-diastolic volume of the ventricles.

A   B   C   D   E

In the following questions, indicate whether the item on the left is greater than (G), the same as (S), or less than (L) the item on the right.

13. Arterial pressure during systole. **G S L** Arterial pressure during diastole.

14. Diastolic arterial pressure in a patient with aortic stenosis. **G S L** Diastolic arterial pressure in a patient with aortic insufficiency.

15. Resting cardiac output while lying down. **G S L** Resting cardiac output while standing up.

16. Cardiac output during exercise. **G S L** Cardiac output at rest.

17. Cardiac output when the heart rate is 65 beats/min. **G S L** Cardiac output when the heart rate is 115 beats/min.

18. Cardiac output when the heart rate is 115 beats/min. **G S L** Cardiac output when the heart rate is 195 beats/min.

19. Increase in O_2 consumption of myocardium produced by increased preload. **G S L** Increase in O_2 consumption of myocardium produced by increased afterload.

20. Duration of systole at a heart rate of 80 beats/min. **G S L** Duration of diastole at a heart rate of 80 beats/min.

21. Duration of systole at a heart rate of 80 beats/min. **G S L** Duration of systole at a heart rate of 180 beats/min.

REFERENCES

Katz AM: *Physiology of the Heart*. Raven Press, 1977.

30   Dynamics of Blood & Lymph Flow

The subject of Chapter 30 is the blood vessels and lymphatics and the movement of fluids through them. The functional anatomy of the arteries, arterioles, capillaries, venules, veins, and lymphatic vessels is summarized, and the principles and forces that govern pressure and flow in them are reviewed. The capillary bed, its importance, and the factors affecting exchange across capillary walls are considered in detail.

OBJECTIVES

The material in the chapter should help students to—

- Describe in relative terms the diameter, wall thickness, and total cross-sectional area of the aorta, smaller arteries, arterioles, capillaries, venules, and veins.

- Describe the relationship between flow, pressure, and resistance in the vascular system.

- Describe the microscopic anatomy of the capillary wall, and relate the morphology to the mechanisms by which substances cross the capillary wall.

- List and assess the methods commonly available for measuring blood flow.

- Define laminar flow and critical closing pressure.

- State the Poiseuille-Hagen formula for flow in blood vessels, and explain on the basis of this formula why the radius of a vessel is such an important determinant of flow.

- Define the law of Laplace, and list 3 examples of its operation in the body.

- Describe in detail how blood pressure in humans is measured by the auscultatory method and the palpation method.

- List the Starling forces that determine the net movement of fluid across the capillary wall, and define flow-limited exchange and diffusion-limited exchange.

- List the factors responsible for the movement of blood in the great veins.

GENERAL QUESTIONS

1. What is a Windkessel? Why are the aorta and large arteries called Windkessel vessels? What is the physiologic significance of the Windkessel effect?

2. What is Bernoulli's principle? Discuss its significance in cardiovascular physiology.

3. A young man has a blood pressure of 130/72 mm Hg. What is his pulse

pressure? What is his mean arterial pressure? How did you calculate each of these pressures?

4. The pressure in a large artery is 60,000 dynes/cm², and the radius of the artery is 0.5 cm. What is the wall tension in the artery? What would the wall tension be if the diameter were 1 cm?

5. Define and explain the sounds of Korotkow.

6. How would you measure the pressure in a capillary?

7. Compare and contrast fenestrated and nonfenestrated capillaries in terms of morphology and function.

8. Where in the cardiovascular system is turbulent flow normally found? What factors make flow change from streamline to turbulent?

9. Discuss the effects of gravity on the circulation.

10. What factors affect central venous pressure?

MULTIPLE-CHOICE QUESTIONS

In the following questions, select the single best answer.

Questions 1–4 refer to Table 30–1.

One hind limb of an anesthetized experimental animal is denervated and attached to instruments so that it is possible to measure the pressure at the arterial and venous ends of the capillaries, the weight of the limb, its blood flow, and the arteriovenous oxygen difference across it. Various substances are then administered.

Table 30–1.

| Pattern | Pressure at | | Weight of Limb | Blood Flow in Limb | Arteriovenous Oxygen Difference |
	Arterial End of Capillary	Venous End of Capillary			
a.	—	—	↑	—	—
b.	—	↑	—	↓	—
c.	↓	↓	—	↓	↑
d.	↑	↑	↑	↑	↓
e.	↑	↑	↑	↓	—

— = no change; ↑ = increase; ↓ = decrease.

1. Which pattern of changes in Table 30–1 would be observed if the substance produced arteriolar dilation?

a   b   c   d   e

2. Which pattern would be observed if the substance produced increased capillary permeability?

a   b   c   d   e

3. Which pattern would be observed if the substance produced constriction of the veins?

a   b   c   d   e

4. Which pattern would be observed if the substance produced a decrease in systemic blood pressure?

a   b   c   d   e

5. Which of the following vessels has the greatest cross-sectional area?
   a. the aorta.
   b. an arteriole.
   c. the inferior vena cava.
   d. the thoracic duct.
   e. the pulmonary artery.

6. In which of the following lists of blood vessels is the sequence of vessels arranged from highest to lowest *total* cross-sectional area in the body?
   a. arteries, arterioles, capillaries, veins.
   b. arterioles, capillaries, arteries, veins.
   c. capillaries, arterioles, veins, arteries.
   d. veins, capillaries, arterioles, arteries.
   e. arteries, veins, arterioles, capillaries.

7. Lymph flow from the foot is—
   a. increased when an individual rises from the supine to the standing position.
   b. increased by massaging the foot.
   c. increased when capillary permeability is decreased.
   d. decreased when the valves of the leg veins are incompetent.
   e. decreased by exercise.

8. Movement of fluid from a capillary into tissues is increased by—
   a. a fall in arterial pressure.
   b. a rise in venous pressure.
   c. a rise in plasma oncotic pressure.
   d. constriction of the arterioles.
   e. closure of precapillary sphincters.

9. The mean pressure in the artery supplying a given organ is 100 mm Hg, the mean pressure in the vein draining it is 10 mm Hg, and the blood flow through the organ is 600 mL/min. What is the vascular resistance across the organ?
   a. 0.1 R unit.
   b. 0.9 R unit.
   c. 1 R unit.
   d. 9 R units.
   e. 11 R units.

In the following questions, one or more than one of the answers may be correct. Select—

A if (1), (2), and (3) are correct;
B if (1) and (3) are correct;
C if (2) and (4) are correct;
D if only (4) is correct; and
E if all are correct.

10. Which of the following vessels contain valves?
    (1) lymphatics.
    (2) veins from the abdominal viscera.

(3) limb veins.
(4) arterioles.

A   B   C   D   E

11. Which of the following help to move lymph toward the heart?
(1) contractions of the walls of large lymphatic ducts.
(2) contractions of skeletal muscle.
(3) pulsation of neighboring arteries.
(4) rapid flow of blood in large veins.

A   B   C   D   E

12. The velocity of blood flow—
(1) is higher in the capillaries, because the total cross-sectional area of the capillaries is large.
(2) is lower in the veins than in the venules.
(3) falls to zero in the descending aorta during diastole.
(4) is increased in a constricted area of a blood vessel.

A   B   C   D   E

13. As blood passes through the systemic capillaries—
(1) its hematocrit increases.
(2) its hemoglobin dissociation curve shifts to the right.
(3) its protein content increases.
(4) its pH decreases.

A   B   C   D   E

14. The pressure in the radial artery is determined by—
(1) pressure in the brachial vein.
(2) the force with which the left ventricle contracts.
(3) pressure in the hepatic portal vein.
(4) the rate of discharge in sympathetic nerve fibers to the arm.

A   B   C   D   E

15. Arterioles in which of the following organs dilate when intraluminal pressure is reduced?
(1) the brain.
(2) the kidneys.
(3) skeletal muscle.
(4) the heart.

A   B   C   D   E

In the following questions, indicate whether the item on the left is greater than **(G)**, the same as **(S)**, or less than **(L)** the item on the right.

16. Resistance of an arteriole   **G S L**   Resistance of an arteriole
with a diameter of 10 μm.   with a diameter of 7 μm.

17. Hydrostatic pressure in venules at the level of the heart.   G   S   L   Hydrostatic pressure in veins at the level of the heart.

18. Permeability of capillaries in skeletal muscle.   G   S   L   Permeability of capillaries in the cerebral cortex.

19. Permeability of capillaries in the lungs.   G   S   L   Permeability of capillaries in the liver.

20. Pressure in the veins in the foot of a man who is standing quietly.   G   S   L   Pressure in the veins in the neck of a man who is standing quietly.

21. Mean arterial pressure in a normal man.   G   S   L   Mean pulse pressure in a normal man.

22. Pressure in capacitance vessels.   G   S   L   Pressure in resistance vessels.

23. Interstitial fluid volume in exercising muscle.   G   S   L   Interstitial fluid volume in resting muscle.

24. Wall thickness in a capillary.   G   S   L   Wall thickness in the aorta.

25. Distending pressure at a point where an artery is narrowed.   G   S   L   Distending pressure proximal to such a constriction.

REFERENCES

Berne RM, Levy MN: *Cardiovascular Physiology*, 5th ed. Mosby, 1986.

Cardiovascular Regulatory Mechanisms

31

Chapter 31 is concerned with the local mechanisms, the circulating humoral agents, and the neural mechanisms that act together to regulate the cardiovascular system. These mechanisms maintain blood pressure and blood flow to the various organs at rest and adjust cardiac output so that pressure and flow are appropriate to meet the needs of various emergencies. The humoral agents include norepinephrine, epinephrine, histamine, neuropeptide Y, atrial natriuretic peptide (ANP), kinins, angiotensin II, and vasopressin. The neurons and pathways in the medulla oblongata that are concerned with regulating sympathetic output to the blood vessels and heart are reviewed, as well as the parasympathetic output to the heart. The structure and function of the arterial baroreceptors and baroreflex regulation of blood pressure are considered in detail.

OBJECTIVES

The material in the chapter should help students to—

- Define autoregulation, discuss its role in physiology, and summarize the theories that have been advanced to explain its occurrence.

- List the important vasodilator metabolites, and discuss their role in the regulation of tissue perfusion.

- List the principal humoral factors in the circulation that affect blood pressure, and comment on the physiologic role of each.

- List the important kinins in the body, and outline the ways they are synthesized and metabolized.

- Describe the innervation of arteries, arterioles, venules, and veins.

- List the factors that affect the caliber of arterioles, and describe the action of each.

- Describe the possible role of peptide cotransmitters in cardiovascular regulation.

- Outline the neural paths that affect heart rate, including the receptors, the pathways, the nuclei involved in central integration.

- Describe the direct effects of CO_2 and hypoxia on the vasomotor areas in the medulla oblongata.

- Describe the sympathetic vasodilator system, and discuss its possible role in the physiologic regulation of the cardiovascular system.

GENERAL QUESTIONS

1. Why does increased sympathetic activity cause an increase in right atrial pressure? What are the mechanisms and pathways involved?

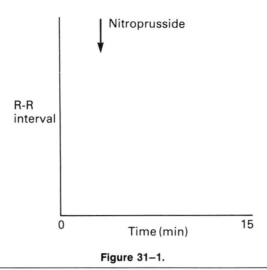

Figure 31-1.

2. Using the axes in Fig 31–1, plot R–R interval against time before and after intravenous injection of the vasodilator nitroprusside at the arrow.

3. What are the mechanisms by which a drug that inhibits angiotensin converting enzyme lowers blood pressure?

4. What is the function of the carotid and aortic baroreceptors? How does their function compare to that of the baroreceptors in the atria and great veins?

5. What hormones affect the heart rate, and how do they produce their effects?

6. What is the Bezold-Jarisch reflex (coronary chemoreflex)?

7. What peptides are found in nerves innervating blood vessels? What is the function of these peptides?

8. When a stimulating electrode is inserted into the medulla oblongata, in which medullary structures would you expect stimulation to increase blood pressure? In which structures would you expect stimulation to decrease blood pressure? Explain your answers.

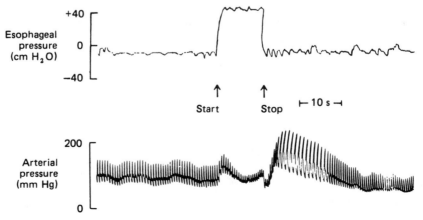

Figure 31-2. Diagram of the response to straining (the Valsalva maneuver) in a normal man, recorded with a needle in the brachial artery. (Courtesy of M McIlroy.)

9. What are the effects of baroreceptor denervation?

10. Fig 31–2 shows the effect of the Valsalva maneuver (straining against a closed glottis) on the blood pressure and heart rate. Why does the blood pressure decline during straining? Why does the blood pressure increase greatly after straining is over? Why does the heart rate slow for 5–10 seconds after the end of straining?

11. Discuss the role of the endothelium in the regulation of vascular contraction and relaxation.

MULTIPLE-CHOICE QUESTIONS

In the following questions, select the single best answer.

1. Which of the following are *incorrectly* paired?
   a. angiotensin converting enzyme : kinin metabolism.
   b. kallikrein : plasma renin activity.
   c. increased pressure in the carotid sinuses : decreased heart rate.
   d. increased pressure in the carotid sinuses : decreased sympathetic discharge to arterioles.
   e. local action of CO_2 in tissues : vasoconstriction in the tissues.

2. Which of the following would *not* be expected to increase heart rate?
   a. stimulation of sympathetic nerves to the heart.
   b. thyrotoxicosis (chronic excess secretion of thyroid hormones).
   c. bilateral section of the ninth and tenth cranial nerves as they enter the skull.
   d. inspiration.
   e. injection of neuropeptide Y.

3. Increased activity of the sympathetic vasodilator system—
   a. has no effect on blood flow in skeletal muscle.
   b. is associated with decreased adrenal medullary secretion.
   c. may shunt blood through arteriovenous anastomoses.
   d. may be responsible for the increase in blood flow in actively exercising muscle once exercise has begun.
   e. produces increased release of epinephrine instead of norepinephrine from postganglionic neurons.

4. When a pheochromocytoma (tumor of the adrenal medulla) suddenly discharges a large amount of epinephrine into the circulation, the patient's heart rate would be expected to—
   a. increase because the increase in blood pressure stimulates the carotid and aortic baroreceptors.
   b. increase because epinephrine has a direct chronotropic effect on the heart.
   c. increase because of increased tonic parasympathetic discharge to the heart.
   d. decrease because the increase in blood pressure stimulates the carotid and aortic chemoreceptors.
   e. decrease because of increased tonic parasympathetic discharge to the heart.

5. Vasopressin secretion is increased by—
   a. increased pressure in the right ventricle.
   b. decreased pressure in the right ventricle.
   c. increased pressure in the right atrium.
   d. decreased pressure in the right atrium.
   e. increased pressure in the aorta.

In the following questions, one or more than one of the answers may be correct. Select—

        A if (1), (2), and (3) are correct;
        B if (1) and (3) are correct;
        C if (2) and (4) are correct;
        D if only (4) is correct; and
        E if all are correct.

6. Activity in sympathetic nerves to the heart is decreased—
    (1) during excitement.
    (2) when blood pressure is suddenly increased by infusion of norepinephrine.
    (3) during hypoxia.
    (4) immediately after lying down from a standing position.

            A   **B**   **C**   **D**   **E**

7. Blood pressure rises with—
    (1) stimulation of the nerves from the aortic bodies.
    (2) stimulation of pain fibers from the abdomen.
    (3) an increase in cardiac output.
    (4) stimulation of the nerves from the carotid sinus.

            A   **B**   **C**   **D**   **E**

8. Blood pressure falls with—
    (1) low oxygen tension in the medulla oblongata.
    (2) rage.
    (3) occlusion of the carotid arteries in the neck.
    (4) stimulation of the vagus nerves that innervate the atria and great veins.

            A   **B**   **C**   **D**   **E**

9. Blood flow is increased by the local action of—
    (1) potassium ions.
    (2) carbon dioxide.
    (3) hydrogen ions.
    (4) histamine.

            A   **B**   **C**   **D**   **E**

10. Catecholamines acting on α-adrenergic receptors—
    (1) increase the contractility of cardiac muscle.
    (2) increase the rate of discharge of the sinoatrial node.
    (3) increase cardiac output.
    (4) constrict coronary arteries by a direct action on these blood vessels.

            A   **B**   **C**   **D**   **E**

In the following questions, indicate whether the item on the left is greater than **(G)**, the same as **(S)**, or less than **(L)** the item on the right.

11. Capillary permeability at the site of subcutaneous injection of histamine.   **G S L**   Capillary permeability at the site of subcutaneous injection of epinephrine.

12. Blood pressure following intravenous injection of histamine.   **G S L**   Blood pressure following intravenous injection of epinephrine.

13. Rate of discharge in a nerve from the aortic arch baroreceptors when the blood pressure is suddenly raised to 180/110 mm Hg.   **G S L**   Rate of discharge in a nerve from the aortic arch baroreceptors when the blood pressure is suddenly lowered to 95/55 mm Hg.

14. Rate of discharge in vaso-motor nerves when the blood pressure is 110/60 mm Hg in a normal individual.   **G S L**   Rate of discharge in vaso-motor nerves when the blood pressure is lowered to 110/60 mm Hg in an individual with chronic hypertension.

15. Mean diameter of arterioles after section of carotid sinus nerves.   **G S L**   Mean diameter of arterioles after stimulation of carotid sinus nerves.

16. Heart rate during inspiration.   **G S L**   Heart rate during expiration.

17. Capacity of veins of an individual standing erect.   **G S L**   Capacity of veins of an individual lying quietly in bed.

18. Heart rate at rest in a transplanted human heart.   **G S L**   Heart rate at rest in a normal human heart.

REFERENCES

Shepherd JT, Vanhoutte PM: *The Human Cardiovascular System*. Raven Press, 1979.

Chapter 32 describes special aspects of the circulation of the brain, heart, splanchnic region, skin, and placenta. The cerebral circulation and its regulation are considered, along with the nature of cerebrospinal fluid and the characteristics of the blood-brain barrier. The coronary circulation and its abnormalities are reviewed. The distribution of blood flow in the various abdominal viscera and the unique features of the hepatic circulation are analyzed. The vascular responses of the skin to injury are described, and the circulation of the placenta and the circulatory alterations that occur in the fetus at the time of birth are considered.

OBJECTIVES

The material in the chapter should help students to—

- Give approximate values for relative blood flow (mL/unit weight/min) and absolute blood flow (mL/min) at rest in various major organs of the body.

- List the unique gross and microscopic aspects of the circulation of the brain.

- List the major ways in which the composition of cerebrospinal fluid differs from that of plasma.

- Describe the formation, absorption, and functions of cerebrospinal fluid.

- Outline the characteristics of the blood-brain barrier, and comment on its importance in clinical medicine.

- Describe the circumventricular organs, and list their general functions.

- Discuss the relationship between local neural activity and local blood flow in the brain.

- List the main energy sources of the brain.

- Summarize the main anatomic features of the coronary circulation.

- List the chemical and neural factors that regulate the coronary circulation, and describe the role of each.

- Outline the unique features of the circulation of the liver and the splanchnic bed, and comment on the reservoir function of the splanchnic circulation.

- Describe the triple response produced by firmly stroking the skin, and explain each of its components.

- Describe the operation of the placenta as the "fetal lung."

- Diagram the circulation of the fetus before birth, and list the changes that occur in it at birth.

- Describe the differences between fetal and adult hemoglobin.

GENERAL QUESTIONS

1. Using the axes of Fig 32–1, plot the curve relating brain blood flow to arterial pressure. Explain the shape of the curve and the underlying physiologic processes responsible for it.

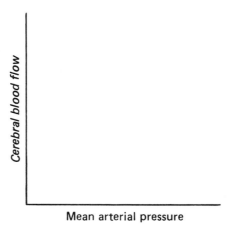

Mean arterial pressure

Figure 32–1.

2. What is the axon reflex? What is the evidence for its existence? What neurotransmitter is probably involved?

3. What is reactive hyperemia? Where does it occur?

4. What is a neurohemal organ? Where are the 2 principal neurohemal organs in the body?

5. How can positron emission tomography be used to study the metabolism and blood flow of the brain?

6. Discuss the use of radionuclides in studying the coronary circulation.

7. What is the mechanism that increases blood flow to the intestine after meals?

8. What is the Monroe-Kellie doctrine, and what are its physiologic consequences?

9. What is a hepatic acinus? Why is the acinar organization of the liver important, and what is its physiologic significance?

MULTIPLE-CHOICE QUESTIONS

In the following questions, select the single best answer. The following 5 questions refer to Fig 32–2.

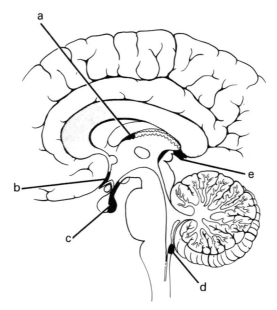

Figure 32–2. Sagittal view of human brain.

1. Main site at which hypothalamic hormones leave the brain

a   b   c   d   e

2. Main site at which vasopressin acts to decrease cardiac output

a   b   c   d   e

3. Chemoreceptor area that is a trigger zone for vomiting

a   b   c   d   e

4. Main site where changes in plasma osmolality act to alter vasopressin secretion

a   b   c   d   e

5. Main site at which angiotensin II acts to increase water intake

a   b   c   d   e

6. Which of the following normally has the lowest blood P_{O_2}?
   a. maternal artery.
   b. maternal uterine vein.
   c. maternal femoral vein.
   d. umbilical artery.
   e. umbilical vein.

7. The pressure differential between the heart and the aorta is least in the—
   a. left ventricle during systole.
   b. left ventricle during diastole.
   c. right ventricle during systole.
   d. right atrium during systole.
   e. left atrium during systole.

8. Injection of tissue plasminogen activator (TPA) would probably be most beneficial—
  a. after at least 1 year of uncomplicated recovery following occlusion of a coronary artery.
  b. after at least 2 months of rest and recuperation following occlusion of a coronary artery.
  c. during the second week after occlusion of a coronary artery.
  d. during the second day after occlusion of a coronary artery.
  e. during the second hour after occlusion of a coronary artery.

9. Which of the following organs has the greatest blood flow per 100 g of tissue?
  a. the brain.
  b. heart muscle.
  c. the skin.
  d. the liver.
  e. the kidneys.

10. Which of the following organs has the most permeable capillaries?
  a. brain.
  b. posterior pituitary gland.
  c. liver.
  d. small intestine.
  e. kidneys.

In the following questions, one or more than one of the answers may be correct. Select—

  A if (1), (2), and (3) are correct;
  B if (1) and (3) are correct;
  C if (2) and (4) are correct;
  D if only (4) is correct; and
  E if all are correct.

11. Cerebrospinal fluid—
  (1) contains very little glucose.
  (2) is a major source of nutritive substances for the brain.
  (3) is a major pathway for delivery of hypothalamic hypophyseotropic hormones to the anterior pituitary gland.
  (4) contains less than 1% as much protein as plasma.

  A   B   C   D   E

12. Cerebrospinal fluid is produced in the—
  (1) arachnoid villi.
  (2) choroid plexus.
  (3) foramens of Monroe.
  (4) spaces around cerebral vessels.

  A   B   C   D   E

13. The coronary arteries are—
  (1) dilated by adenosine.
  (2) constricted by a direct effect of norepinephrine.
  (3) dilated by an indirect effect of norepinephrine.
  (4) constricted by hypoxia.

  A   B   C   D   E

14. Arterioles in the skin dilate when exposed to—
   (1) increased body temperature.
   (2) epinephrine.
   (3) bradykinin.
   (4) vasopressin.

A   B   C   D   E

15. Which of the following are important blood reservoirs that can help provide extra circulating arterial blood during emergencies?
   (1) the skin.
   (2) the splanchnic region.
   (3) the lungs.
   (4) skeletal muscle.

A   B   C   D   E

In the following questions, indicate whether the item on the left is greater than **(G)**, the same as **(S)**, or less than **(L)** the item on the right.

16. Amount of water actively transported into the CSF.   **G S L**   Amount of water actively transported out of the CSF.

17. Total O_2 content of fetal blood when the P_{O_2} is 80 mm Hg.   **G S L**   Total O_2 content of adult blood when the P_{O_2} is 80 mm Hg.

18. Blood flow in the motor cortex of the brain at rest.   **G S L**   Blood flow in the motor cortex of the brain when the fists are clenched.

19. Local capillary permeability after subcutaneous injection of antagonist to substance P.   **G S L**   Local capillary permeability after subcutaneous injection of substance P.

20. Blood flow in the endocardial portion of the myocardium during systole.   **G S L**   Blood flow in the epicardial portion of the myocardium during systole.

21. Blood flow in the left coronary artery during systole.   **G S L**   Blood flow in the left coronary artery during diastole.

22. Blood flow in the right coronary artery during systole.   **G S L**   Blood flow in the right coronary artery during diastole.

23. Amount of blood brought to the liver by the hepatic artery.   **G S L**   Amount of blood brought to the liver by the portal vein.

24. Blood flow through the ductus arteriosus before birth.   **G S L**   Blood flow through the ductus arteriosus after birth.

25. Blood flow through the   **G S L**   Blood flow through the
    lungs before birth.   lungs after birth.

26. Blood flow through the   **G S L**   Blood flow through the
    foramen ovale before birth.   foramen ovale after birth.

REFERENCES

Fishman RA: *Cerebrospinal Fluid in Diseases of the Nervous System*. Saunders, 1980.
Shepherd JT, Abboud FM (editors): Regulation of circulation to individual vascular beds. In: *The Cardiovascular System*. Section I, Vol 3 of: *Handbook of Physiology*. American Physiological Society, 1983.

33

Cardiovascular Homeostasis in Health & Disease

Chapter 33 summarizes the cardiovascular adjustments that occur on assuming the upright position, during space flight, and during exercise. It also describes the pathophysiology of shock, syncope, hypertension, and heart failure and the cardiovascular compensations that occur in these conditions.

OBJECTIVES

The material in the chapter should help students to—

- Outline the compensatory mechanisms that maintain blood pressure on rising from the supine to the standing position.

- Describe the effects of positive and negative gravitational acceleration (g) on the body.

- Summarize the effect of zero gravity in space flight on the cardiovascular system and other systems in the body.

- Compare the composition of blood from exercising skeletal muscle with that of blood from resting muscle, and describe how the composition of blood in the systemic circulation changes during exercise.

- Outline the main beneficial effects of athletic training on cardiovascular performance during exercise.

- Define shock, name its major causes, and summarize the main abnormalities that occur and the short- and long-term compensatory mechanisms that operate in each.

- List the common causes of fainting.

- List the common causes of high blood pressure in humans, and divide them into those that are now curable and those that are not.

- List the main symptoms of heart failure, and describe how each is produced.

GENERAL QUESTIONS

1. Why do some people faint when they rise from the supine to the standing position?

2. Why is vital capacity reduced in heart failure? What is the effect of posture on this reduction? What treatment would you prescribe to improve the vital capacity?

3. What are the similarities and differences between hypovolemic shock and low-resistance shock?

4. Describe and explain the circulatory changes that occur during exercise.

5. What changes occur in body function during space flights that last 3–20 days? What further changes might be expected to occur with much longer space flights, eg, a trip to Mars?

6. Why are angiotensin converting enzyme inhibitors of value in the treatment of congestive heart failure?

7. Hypertension is a syndrome, not a disease. Discuss this statement and its implications.

8. List 5 ways that chronic hypertension can be produced in experimental animals. What mechanism produces the elevation of blood pressure in each case? Discuss the similarities and differences between each form of experimental hypertension and human hypertension.

9. What is irreversible shock? How can it be prevented?

10. Discuss the physiologic advantages and disadvantages of treating shock by raising the foot of the bed.

11. Mr. Smith, a 55-year-old man who weighs 71 kg, feels tired after relatively mild exertion. On physical examination, he is found to have a moderately intense systolic murmur. During cardiac catheterization, the following data were obtained:

Heart rate (beats/min)	64
Ventilation (L/min)	4.98
O_2 consumption (mL/min)	246
Pulmonary arterial O_2 (mL of O_2/L of blood)	137
Pulmonary venous O_2 (mL of O_2/L of blood)	189
Right ventricular pressure*	27/2
Pulmonary artery pressure*	27/10
Left ventricular pressure*	169/0
Aortic pressure*	108/72

What is Mr. Smith's cardiac cardiac output? What is his stroke volume? What is the most likely diagnosis?

MULTIPLE-CHOICE QUESTIONS

In the following questions, select the single best answer.

Each of the patients in questions 1–5 has a blood pressure of 85/40 mm Hg. In each case, match the condition with the best emergency treatment among the lettered items below.
  a. Injection of dopamine.
  b. Infusion of concentrated human albumin.
  c. Infusion of isotonic saline solution.
  d. Injection of epinephrine.
  e. Infusion of whole blood.

1. Girl stung by a bee

        a   b   c   d   e

---

* Systolic/diastolic pressure (in mm Hg).

2. Man bleeding from a stab wound

a   b   c   d   e

3. Woman burned over 35% of her body in a kitchen fire

a   b   c   d   e

4. Woman who had a myocardial infarction 24 hours previously

a   b   c   d   e

5. Man with severe diarrhea

a   b   c   d   e

6. Plasma renin activity is most likely to be lower than normal—
  a. in congestive heart failure.
  b. in hemorrhagic shock.
  c. in shock due to infection with gram-negative bacteria.
  d. in essential hypertension.
  e. during quiet standing.

7. Which of the following takes longest to return to normal after 1 L of blood is removed from a normal individual?
  a. plasma aldosterone concentration.
  b. blood pressure.
  c. renin secretion.
  d. plasma volume.
  e. number of red blood cells in peripheral blood.

8. Which of the following would probably cause the most severe sustained hypertension?
  a. chronically increased secretion of the adrenal medulla.
  b. chronically increased secretion of the zona fasciculata and zona reticularis of the adrenal cortex.
  c. chronically increased secretion of the zona glomerulosa of the adrenal cortex.
  d. chronically increased secretion of the thyroid gland.
  e. treatment with oral contraceptives.

9. Which of the following is *not* increased during exercise?
  a. respiratory rate.
  b. stroke volume.
  c. heart rate.
  d. total peripheral resistance.
  e. systolic blood pressure.

10. Which of the following are *incorrectly* paired?
  a. loss of blood : hypotension.
  b. negative *g* : blackout.
  c. increased cardiac output : exercise.
  d. renal artery constriction : increased blood pressure.
  e. decreased total peripheral resistance : fainting.

In the following questions, one or more than one of the answers may be correct. Select—

  A if (1), (2), and (3) are correct;
  B if (1) and (3) are correct;
  C if (2) and (4) are correct;
  D if only (4) is correct; and
  E if all are correct.

11. Postural hypotension can be caused by drugs that block peripheral α_1-adrenergic receptors or diseases that damage postganglionic sympathetic neurons. Alternatively, postural hypotension can be caused by degeneration of pathways in the central nervous system that regulate sympathetic outflow. Which of the following would you predict?
    (1) the plasma epinephrine level would not increase upon standing when the defect is in the central nervous system but would increase when postganglionic sympathetic neurons are damaged.
    (2) the plasma norepinephrine level would not increase upon standing when the defect is in the central nervous system but would increase when postganglionic sympathetic neurons are damaged.
    (3) sweating would be decreased when the defect is in the central nervous system and when postganglionic sympathetic neurons are destroyed.
    (4) the heart rate would increase upon standing when the defect is central but not when it is due to blockade of peripheral adrenergic receptors.

    A   B   C   D   E

12. The causes of syncope (fainting) include—
    (1) pressure on the carotid sinus.
    (2) paroxysmal ventricular fibrillation.
    (3) autonomic insufficiency.
    (4) strong emotion.

    A   B   C   D   E

13. The effects of prolonged physical training on the cardiovascular system include—
    (1) increased maximum O_2 extraction by tissues.
    (2) decreased resting heart rate.
    (3) increased heart size.
    (4) increased mean arterial pressure.

    A   B   C   D   E

14. Which of the following occur in hemorrhagic shock?
    (1) increased secretion of vasopressin.
    (2) increased secretion of glucagon.
    (3) increased secretion of aldosterone.
    (4) increased secretion of epinephrine and norepinephrine.

    A   B   C   D   E

15. Which of the following would you expect to increase in a normal individual who stands quietly in the same position for 1 hour?
    (1) hematocrit.
    (2) diameter of the thigh.
    (3) plasma renin activity.
    (4) lymph flow from the legs.

    A   B   C   D   E

In the following questions, indicate whether the item on the left is greater than **(G)**, the same as **(S)**, or less than **(L)** the item on the right.

16. Plasma protein concentra- **G   S   L**   Plasma protein concentra-
    tion at A in Fig 33–1.   tion at B in Fig 33–1.

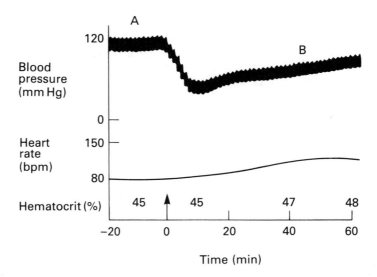

Figure 33–1. Blood pressure, heart rate, and hematocrit of a dog before and after performance of a procedure at time zero.

17. Interstitial fluid volume at **G   S   L**   Interstitital fluid volume at
    A in Fig 33–1.   B in Fig 33–1.

18. Plasma ACTH concentra- **G   S   L**   Plasma ACTH concentra-
    tion at A in Fig 33–1.   tion at B in Fig 33–1.

19. Blood flow to brain during **G   S   L**   Blood flow to brain at rest.
    exercise.

20. Blood flow to kidney dur- **G   S   L**   Blood flow to kidney at
    ing exercise.   rest.

21. Blood flow to skeletal mus- **G   S   L**   Blood flow to skeletal mus-
    cle during exercise.   cle at rest.

22. Stroke volume during iso- **G   S   L**   Stroke volume during iso-
    tonic exercise.   metric exercise.

23. Central blood volume in an **G   S   L**   Central blood volume in an
    astronaut in orbit for 1   astronaut on the ground.
    hour.

24. Blood pressure of a normal **G   S   L**   Blood pressure of the same
    individual when lying   normal individual when
    down.   standing.

25. Cardiac output when lying **G   S   L**   Cardiac output when stand-
    down.   ing.

26. Total peripheral resistance **G   S   L**   Total peripheral resistance
    when lying down.   when standing.

27. Incidence of stroke in hypertension.

G S L

Incidence of stroke in an individual with normal blood pressure.

REFERENCES

Frohlich ED (editor): *Pathophysiology,* 3rd ed. Section I: Circulatory mechanisms, pages 1–118. Lippincott, 1984.

34

Pulmonary Function

Chapter 34 is an analysis of the functions of the respiratory system. A brief summary of the properties of gases is followed by an analysis of the way the lungs and chest operate to produce inspiration and expiration. Gas exchange in the lungs, diffusion capacity for O_2, the special features of the pulmonary circulation, lung defense mechanisms, and the metabolic functions of the lung are also reviewed.

OBJECTIVES

The material in the chapter should help students to—

- Define partial pressure, and calculate the partial pressure of each of the important gases in the atmosphere at sea level.

- Draw a graph of the changes in intrapulmonary and intrapleural pressure and breath volume that occur during inspiration and expiration.

- Outline the anatomy of the air passages, and describe the cells that line them.

- Describe the major factors involved in the regulation of the diameter of the airways.

- List the major muscles involved in respiration, and state the role of each.

- Define tidal volume, inspiratory reserve volume, expiratory reserve volume, and residual volume, and give approximate values for each in a normal adult.

- Define compliance, and give examples of diseases in which it is abnormal.

- Describe the chemical composition and function of surfactant.

- List the factors that determine alveolar ventilation.

- Define diffusion capacity, and compare the diffusion of O_2 with that of CO_2 in the lungs.

- Compare the pulmonary and systemic circulations, listing the main differences between them.

- Describe the metabolic functions of the lungs.

GENERAL QUESTIONS

1. Describe the changes in intrapulmonary and intrapleural pressure that occur during the respiratory cycle. Why do the intrapleural changes lag behind the intrapulmonary changes?

2. Discuss the law of Laplace as it relates to pulmonary function.

3. On the axes in Fig 34–1, draw the line relating relative blood flow in the lungs in the standing position (mid portion of lungs = 100) to vertical distance along the lungs. Explain the differences in blood flow.

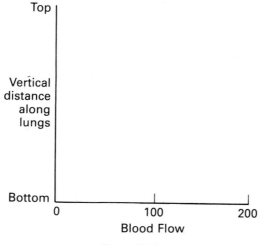

Figure 34–1.

4. What is a hysteresis loop? Give an example in respiratory physiology, and explain its occurrence.

5. What are the functions of pulmonary epithelial cells and pulmonary macrophages?

6. What are the anatomic and physiologic dead spaces? What is the significance of each in terms of normal and abnormal pulmonary function?

7. What are the factors that affect the diffusion capacity of the lungs for O_2?

8. Discuss the airway mechanisms that help protect the lungs from injury and disease.

9. Why is the P_{O_2} of blood in the aorta slightly less than the P_{O_2} of blood in the pulmonary veins?

10. What components contribute to the work of breathing? Discuss the normal and pathologic conditions that cause alterations in each component.

11. From midinspiration, a normal subject takes as deep a breath of 100% O_2 as possible, then exhales steadily while the N_2 content of the expired gas is monitored continuously. A plot of the percentage of N_2 in the expired gas versus the volume expired is shown in Fig 34–2. Explain each of the numbered components of the curve.

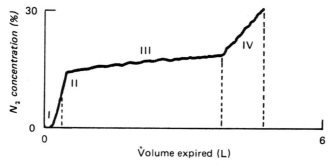

Figure 34–2. Single-breath N_2 curve. The various phases of the curve are indicated by roman numerals. (Modified from Buist AS: New tests to assess lung function: The single-breath nitrogen test. *N Engl J Med* 1975;293:438.)

MULTIPLE-CHOICE QUESTIONS

In the following questions, select the single best answer.

1. On the summit of Mt. Everest, where the barometric pressure is about 250 mm Hg, the partial pressure of O_2 is about—
   a. 0.1 mm Hg.
   b. 0.5 mm Hg.
   c. 5 mm Hg.
   d. 50 mm Hg.
   e. 100 mm Hg.

2. The approximate amount of gas left in the lungs after maximal forced expiration in a normal woman is—
   a. zero.
   b. 0.1 L.
   c. 1.1 L.
   d. 3.1 L.
   e. 4.2 L.

3. The tidal volume in a normal man at rest is about—
   a. 0.5 L.
   b. 1.2 L.
   c. 3.3 L.
   d. 4.8 L.
   e. 6 L.

4. What is the approximate physiologic dead space of a normal 70-kg man breathing through a tube that has a radius of 5 mm and a length of 100 cm?
   a. 150 mL.
   b. 180 mL.
   c. 230 mL.
   d. 280 mL.
   e. 350 mL.

5. Which of the following is responsible for the movement of O_2 from the alveoli into the blood in the pulmonary capillaries?
   a. active transport.
   b. filtration.
   c. secondary active transport.
   d. facilitated diffusion.
   e. passive diffusion.

The following 3 questions refer to Fig 34–3.

6. Which of the labeled curves in Fig 34–3 would you expect to see in a normal individual?

   a   b   c   d   e

7. Which of the labeled curves in Fig 34–3 would you expect to see in a patient with severe pulmonary fibrosis?

   a   b   c   d   e

8. Which of the labeled curves in Fig 34–3 would you expect to see in a patient with advanced emphysema?

   a   b   c   d   e

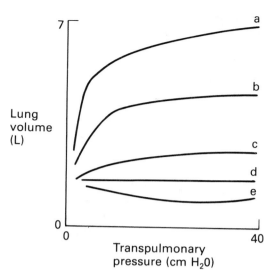

Figure 34–3. Lung volumes at various transpulmonary pressures.

In the following questions, one or more than one of the answers may be correct. Select—

    A if (1), (2), and (3) are correct;
    B if (1) and (3) are correct;
    C if (2) and (4) are correct;
    D if only (4) is correct; and
    E if all are correct.

9. Which of the following cause contraction of bronchial smooth muscle?
    (1) leukotrienes.
    (2) vasoactive intestinal polypeptide (VIP).
    (3) acetylcholine.
    (4) epinephrine.

        A   **B**   **C**   **D**   **E**

10. Surfactant lining the alveoli—
    (1) helps prevent alveolar collapse.
    (2) is decreased in hyaline membrane disease.
    (3) is decreased in the lungs of heavy smokers.
    (4) is a mixture of proteins and lipids.

        A   **B**   **C**   **D**   **E**

11. During inspiration, there is an increase in—
    (1) intra-abdominal pressure.
    (2) venous return to the heart.
    (3) heart rate.
    (4) intrapleural pressure.

        A   **B**   **C**   **D**   **E**

12. Which of the following would be expected to cause a reduction in pulmonary ventilation?
    (1) transection of both phrenic nerves.
    (2) transection of the spinal cord at the first thoracic level.
    (3) a large dose of morphine.
    (4) pulmonary fibrosis.

    A   B   C   D   E

13. In which of the following would it be likely for atelectasis (collapse of pulmonary alveoli) to occur?
    (1) in hyaline membrane disease.
    (2) if there is an open communication between the exterior and the pleural space.
    (3) in chronic respiratory disease with mucus in the bronchi.
    (4) in pulmonary fibrosis.

    A   B   C   D   E

14. Pulmonary diffusing capacity for O_2 is dependent on the—
    (1) P_{O_2} of blood in the pulmonary vein.
    (2) thickness of the alveolar capillary membrane.
    (3) total area of alveoli apposed to capillaries.
    (4) hemoglobin concentration of blood in pulmonary capillaries.

    A   B   C   D   E

15. Airway resistance—
    (1) is increased if the lungs are removed and inflated with saline.
    (2) does not affect the work of breathing.
    (3) is increased in paraplegic patients.
    (4) is increased in asthma.

    A   B   C   D   E

In the following questions, indicate whether the item on the left is greater than (G), the same as (S), or less than (L) the item on the right.

16. Inspiratory reserve volume at rest.   **G S L**   Expiratory reserve volume at rest.

17. Alveolar ventilation when the respiratory rate is 25 breaths/min and the tidal volume is 200 mL.   **G S L**   Alveolar ventilation when the respiratory rate is 10 breaths/min and the tidal volume is 500 mL.

18. Vital capacity in a normal man.   **G S L**   Vital capacity in a normal woman.

19. Amount of gas going to the apexes of the lungs during early inspiration.   **G S L**   Amount of gas going to the bases of the lungs during early inspiration.

20. Partial pressure of O_2 at sea level.   **G S L**   Partial pressure of N_2 at sea level.

21. During quiet breathing at rest, percent of ventilation due to contraction of the diaphragm.   **G S L**   During quiet breathing at rest, percent of ventilation due to contraction of the intercostal muscles.

22. Angiotensin II concentration in blood in pulmonary veins.   **G S L**   Angiotensin II concentration in blood in renal veins.

23. Bradykinin concentration in blood in pulmonary vein.   **G S L**   Bradykinin concentration in blood in inferior vena cava.

24. Amount of smooth muscle in pulmonary arterioles.   **G S L**   Amount of smooth muscle in pulmonary venules.

25. Mean pressure in the pulmonary artery.   **G S L**   Mean pressure in the aorta.

REFERENCES

West JB: *Respiratory Physiology—The Essentials,* 2nd ed. Williams & Wilkins, 1979.

35

Gas Transport Between the Lungs & the Tissues

Chapter 35 describes the flow of O_2 from the lungs to the tissues and the flow of CO_2 from the tissues to the lungs, with emphasis on the physical and chemical mechanisms that greatly augment the ability of the blood to carry O_2 and CO_2. In this context, the binding of O_2 to hemoglobin and the factors modifying O_2 uptake and release are considered, along with buffers and the buffering reactions involved in CO_2 transport.

OBJECTIVES

The material in the chapter should help students to—

- Describe the manner in which O_2 flows "downhill" from the lungs to the tissues and CO_2 flows "downhill" from the tissues to the lungs.

- Describe the reactions of O_2 with hemoglobin.

- List the important factors affecting the affinity of hemoglobin for O_2, and comment on the physiologic significance of each.

- Define myoglobin, and outline its physiologic role.

- Write the Henderson-Hasselbalch equation, and comment on its significance.

- List the principal buffers in blood, interstitial fluid, and intracellular fluid.

- Describe the fate of CO_2 in the blood.

- Draw the CO_2 dissociation curve for arterial (oxygenated) and venous (deoxygenated) blood.

GENERAL QUESTIONS

1. Why is the Cl^- concentration inside red blood cells in venous blood greater than that in red blood cells in arterial blood? Describe the mechanisms responsible for this difference.

2. Why can oxyhemoglobin bind less H^+ than can reduced hemoglobin?

3. What is a blood substitute? Describe the properties of an ideal blood substitute and the potential clinical value of such a substance.

4. What is the Bohr effect? How is it brought about, and what is its physiologic significance?

5. Discuss 2,3-diphosphoglycerate (2,3-DPG). Define P_{50}, and describe how 2,3-DPG affects the P_{50} in blood.

6. What is the role of carbonic anhydrase in red blood cells?

7. Each 100 mL of normal arterial blood contains about 49 mL of CO_2. Describe the distribution of this 49 mL of CO_2 within the red cells and the plasma.

8. What is the oxygen hemoglobin dissociation curve? What factors are responsible for its characteristic shape?

MULTIPLE-CHOICE QUESTIONS

In the following questions, select the single best answer.

1. Most of the CO_2 transported in the blood is—
   a. dissolved in plasma.
   b. in carbamino compounds formed from plasma proteins.
   c. in carbamino compounds formed from hemoglobin.
   d. bound to Cl^-.
   e. in HCO_3^-.

2. Which of the following has the greatest effect on the ability of blood to transport oxygen?
   a. the capacity of the blood to dissolve oxygen.
   b. the amount of hemoglobin in the blood.
   c. the pH of plasma.
   d. the CO_2 content of red blood cells.
   e. the temperature of the blood.

3. Which of the following has the greatest buffering capacity in the interstitial fluid?
   a. the hemoglobin system.
   b. the bicarbonate system.
   c. the phosphate system.
   d. the proteins in the interstitial fluid.
   e. the ammonium-ammonia system.

In the following questions, select—

   A if the item is associated with (a) below,
   B if the item is associated with (b) below,
   C if the item is associated with both (a) and (b), and
   D if the item is associated with neither (a) nor (b).

   (a) Involved in O_2 transport in blood.
   (b) Involved in CO_2 transport in blood.

4. Carbamino compounds   **A B C D**

5. Hemoglobin   **A B C D**

6. Myoglobin   **A B C D**

7. Hydrogen ions   **A B C D**

8. Plasma proteins   **A B C D**

In the following questions, one or more than one of the answers may be correct. Select—

   A if (1), (2), and (3) are correct;
   B if (1) and (3) are correct;

C if (2) and (4) are correct;
D if only (4) is correct; and
E if all are correct.

9. Which of the following are true of the system
$$CO_2 + H_2O \overset{1}{\rightleftharpoons} H_2CO_3 \overset{2}{\rightleftharpoons} H^+ + HCO_3^-?$$
(1) reaction 1 is catalyzed by carbonic anhydrase.
(2) because of reaction 2, the pH of blood declines during breath holding.
(3) reaction 1 occurs in the kidneys.
(4) reaction 1 occurs primarily in plasma.

A   B   C   D   E

10. The flow of O_2 from blood to the tissues is increased by a decrease in—
(1) the 2,3-DPG content of the blood.
(2) blood P_{CO_2}.
(3) plasma Na^+ concentration.
(4) plasma pH.

A   B   C   D   E

11. The concentration of 2,3-DPG in peripheral blood—
(1) determines in part the affinity of hemoglobin for O_2.
(2) increases in chronic hypoxia.
(3) decreases when the H^+ concentration in red cells increases.
(4) is unaffected by exercise.

A   B   C   D   E

In the following questions, indicate whether the item on the left is greater than (G), the same as (S), or less than (L) the item on the right.

12. CO_2 content of arterial blood.   G S L CO_2 content of venous blood.

13. Oxyhemoglobin content of arterial blood.   G S L   Oxyhemoglobin content of venous blood.

14. pH of arterial blood.   G S L   pH of venous blood.

15. Amount of CO_2 in red blood cells when carbonic anhydrase is inhibited.   G S L   Amount of CO_2 in red blood cells when carbonic anhydrase is stimulated.

16. pK′ of phosphate system ($H_2PO_4^- \rightleftharpoons H^+ + HPO_4^{2-}$).   G S L   pK′ of bicarbonate system ($H_2CO_3 \rightleftharpoons H^+ + HCO_3^-$).

17. Affinity of the first heme in the hemoglobin molecule for O_2.   G S L   Affinity of the fourth heme in the hemoglobin molecule for O_2.

18. Affinity of fetal hemoglobin (hemoglobin F) for O_2.   G S L   Affinity of adult hemoglobin (hemoglobin A) for O_2.

19. Number of osmotically active particles in red blood cells in arterial blood.

G S L Number of osmotically active particles in red blood cells in venous blood.

REFERENCES

Davenport HW: *The ABC of Acid-Base Chemistry*, 6th ed. Univ of Chicago Press, 1974.

36

Regulation of Respiration

Chapter 36 reviews the mechanisms that regulate respiration. The regions in the pons and medulla oblongata that control respiration and their interactions are analyzed. The carotid and aortic bodies and the respiratory receptors on the ventral surface of the medulla oblongata are discussed, and their individual roles and interactions in the chemical control of breathing are summarized. Respiratory responses to nonchemical influences such as joint movement and stimulation of stretch and irritant receptors in the lungs are also reviewed.

OBJECTIVES

The material in the chapter should help students to—

- Locate and comment on the function of the dorsal and ventral groups of respiratory neurons, the pneumotaxic center, and the apneustic center in the brain stem.

- List the effects on respiration that are mediated by the vagus nerves.

- List the specific functions of the respiratory receptors in the carotid body, the aortic body, and the ventral surface of the medulla oblongata.

- Describe and explain the ventilatory responses to increased CO_2 concentrations in the inspired air.

- Describe and explain the ventilatory responses to decreased O_2 concentrations in the inspired air.

- List the main nonchemical factors that influence respiration, and outline the effect of each.

- Describe the effect of sleep on respiration.

GENERAL QUESTIONS

1. Ventilation is sensitive to small changes in the CO_2 concentration of inspired air. How do ventilatory responses to these changes occur?

2. Discuss the Hering-Breuer reflexes.

3. What role does the blood-brain barrier play in the regulation of respiration?

4. What is Ondine's curse? Discuss its pathophysiology and the conditions in which it occurs.

5. What is sleep apnea? Discuss its relation to sudden infant death syndrome.

6. What is the breaking point? Discuss the factors that affect it.

MULTIPLE-CHOICE QUESTIONS

In the following questions, select the single best answer.

1. Which of the following discharge spontaneously during quiet breathing?
    a. inspiratory neurons.
    b. motor neurons to respiratory muscles.
    c. neurons in the apneustic center.
    d. expiratory neurons.
    e. stretch receptors in the lungs.

2. Intravenous lactic acid increases ventilation. The receptors responsible for this effect are located in—
    a. the medulla oblongata.
    b. the carotid bodies.
    c. the lung parenchyma.
    d. the aortic baroreceptors.
    e. the trachea and large bronchi.

3. Spontaneous respiration ceases after—
    a. transection of the brain stem above the pons.
    b. transection of the brain stem at the caudal end of the medulla.
    c. bilateral vagotomy.
    d. bilateral vagotomy combined with transection of the brain stem at the superior border of the pons.
    e. transection of the spinal cord at the level of the first thoracic segment.

4. Respiration is subject to chemical and nonchemical control. Variations in which of the following do *not* regularly exert a nonchemical effect on breathing?
    a. afferents from tendons.
    b. afferents from the cerebral cortex.
    c. afferents from the heart.
    d. activity of J receptors in the lungs.
    e. activity of irritant receptors in the trachea.

5. The following physiologic events that occur in the body are listed in random order:
    1. decreased CSF pH.
    2. increased arterial P_{CO_2}.
    3. increased CSF P_{CO_2}.
    4. stimulation of medullary chemoreceptors.
    5. increased alveolar P_{CO_2}.

    What is the usual sequence in which they occur when they affect respiration?
    a. 1, 2, 3, 4, 5.
    b. 4, 1, 3, 2, 5.
    c. 3, 4, 5, 1, 2.
    d. 5, 2, 3, 1, 4.
    e. 5, 3, 2, 4, 1.

In the following questions, one or more than one of the answers may be correct. Select—

> **A** if (1), (2), and (3) are correct;
> **B** if (1) and (3) are correct;
> **C** if (2) and (4) are correct;
> **D** if only (4) is correct; and
> **E** if all are correct.

6. Injection of a drug that stimulates the carotid bodies would be expected to cause—
(1) a decrease in the pH of arterial blood.
(2) an increase in the P_{O_2} of arterial blood.
(3) an increase in the HCO_3^- concentration of arterial blood.
(4) a decrease in the H^+ concentration of arterial blood.

A   B   C   D   E

7. Stimulation of the central (proximal) end of a cut vagus nerve would be expected to—
(1) decrease the blood pressure.
(2) increase the heart rate.
(3) inhibit inspiration.
(4) facilitate inspiration.

A   B   C   D   E

8. The respiratory center—
(1) sends out regular bursts of impulses to expiratory muscles during quiet respiration.
(2) is unaffected by stimulation of pain receptors.
(3) is located in the pons and midbrain.
(4) sends out regular bursts of impulses to inspiratory muscles during quiet respiration.

A   B   C   D   E

9. The stimulation of respiration is relatively slight when the P_{O_2} in the inspired air is reduced from 160 mm Hg to 60 mm Hg, because—
(1) there is a compensatory increase in alveolar P_{O_2}.
(2) there is a slight increase in the pH of arterial blood that tends to inhibit respiration.
(3) the threshold arterial P_{O_2} at which the chemoreceptors are stimulated is 50 mm Hg.
(4) there is a decrease in alveolar P_{CO_2}.

A   B   C   D   E

10. Variations in which of the following components of blood or cerebrospinal fluid affect respiration?
(1) arterial K^+ concentration.
(2) arterial H^+ concentration.
(3) arterial Na^+ concentration.
(4) cerebrospinal fluid CO_2 concentration.

A   B   C   D   E

In the following questions, indicate whether the item on the left is greater than (G), the same as (S), or less than (L) the item on the right.

11. Effect of Hering-Breuer inflation reflex when the tidal volume is 500 mL.

    G S L   Effect of Hering-Breuer inflation reflex when the tidal volume is 1500 mL.

12. Alveolar ventilation when the plasma pH is 7.5.

    G S L   Alveolar ventilation when the plasma pH is 7.2.

13. Alveolar ventilation when the cerebrospinal fluid pH is 7.5.

    G S L   Alveolar ventilation when the cerebrospinal fluid pH is 7.2.

14. Respiratory rate when the brain stem is transected in the inferior portion of the pons when the vagus nerves are intact.

    G S L   Respiratory rate when the brain stem is transected in the inferior portion of the pons after vagotomy.

15. Proportion of the alteration in ventilation produced by changes in arterial P_{CO_2} that is due to medullary chemoreceptors.

    G S L   Proportion of the alteration in ventilation produced by changes in arterial P_{CO_2} that is due to peripheral chemoreceptors.

16. Blood flow per gram of tissue to carotid bodies.

    G S L   Blood flow per gram of tissue to brain.

17. Discharge rate in a nerve from the carotid body when the arterial P_{O_2} is 90 mm Hg.

    G S L   Discharge rate in a nerve from the carotid body when the arterial P_{O_2} is 9 mm Hg.

18. Ventilation when the P_{CO_2} of inspired air is 0.3 mm Hg.

    G S L   Ventilation when the P_{CO_2} of inspired air is 3 mm Hg.

19. Ventilation in a sleeping subject when the P_{CO_2} of inspired air is 50 mm Hg.

    G S L   Ventilation in a waking subject when the P_{CO_2} of inspired air is 50 mm Hg.

20. Ventilation during passive movement of the legs.

    G S L   Ventilation during active movement of the legs.

REFERENCES

Mines AH: *Respiratory Physiology,* 2nd ed. Raven Press, 1986.

37

Respiratory Adjustments in Health & Disease

This chapter considers the changes in respiration that occur with exercise, during exposure to altitude and other forms of hypoxia, during hypercapnia and hypocapnia, and with increased barometric pressure. The 4 principal forms of hypoxia are described, and the major causes of each are listed. Respiratory failure, drowning, and artificial respiration are also discussed.

OBJECTIVES

The material in the chapter should help students to—

- Describe the effects of exercise on ventilation and O_2 exchange in the tissues.

- Define hypoxia, and list its 4 principal forms.

- Describe the effects on the body as one ascends to higher altitudes.

- Define acclimatization to altitude, and summarize the changes that occur in the body with prolonged residence at high altitude.

- Define and give examples of ventilation-perfusion imbalance.

- List and explain the effects of carbon monoxide on the body.

- List and explain the adverse effects of excess O_2.

- Describe the effects of hypercapnia and hypocapnia, and give examples of conditions that can cause them.

- Define periodic breathing, and explain its occurrence in various disease states.

- Describe in detail the technique of mouth-to-mouth resuscitation, and explain how it works to maintain life.

GENERAL QUESTIONS

1. By means of a diagram, illustrate the relationship between partial pressure of O_2 in blood and percent saturation of hemoglobin at rest, and the changes produced in this relationship by exercise.

2. What is the oxygen debt mechanism? How is an oxygen debt measured? What is the value of the mechanism to the individual?

3. Discuss fatigue from the point of view of its causes, prevention, and physiologic significance.

4. Cheyne-Stokes respiration is characterized by periods of hyperventilation alternating with periods of apnea. What causes it? Why?

5. "Chronic hypoxia is a more potent stimulus to ventilation than acute hypoxia, whereas chronic hypercapnia is a weaker stimulus to ventilation than acute hypercapnia." Discuss and explain this statement.

6. If a healthy young adult accidentally aspirates a piece of food that totally occludes the bronchus to a large part of one lobe of the right lung, what will be the immediate consequences, what compensatory responses will be observed, and how do the compensatory responses minimize the effects of the obstruction?

7. In patients with severe respiratory failure who are hypercapnic and hypoxic, administration of O_2 may stop respiration and even cause death if artificial respiration is not instituted. Why?

8. What is the difference between hyperventilation and dyspnea?

9. Emphysema is a serious, progressive, and common respiratory disease. Discuss its causes, its pathophysiology, and its treatment in terms of correcting the physiologic abnormalities it produces.

10. Diving not only is performed by professionals but, with the development of scuba equipment, has become a popular sport. What medical problems can be produced by diving, and how would you treat each of them?

MULTIPLE-CHOICE QUESTIONS

In the following questions, select the single best answer.

1. In which of the following conditions is CO_2 retention most likely to occur?
   a. climbing a high mountain.
   b. ventilatory failure.
   c. carbon monoxide poisoning.
   d. lung failure.
   e. hysterical hyperventilation.

The next 5 questions refer to Table 37–1.

Table 37–1. P_{O_2} (mm Hg).

	Superior Vena Cava	Right Ventricle	Alveolar Gas	Left Ventricle
a.	45	40	104	94
b.	40	35	105	94
c.	40	35	60	55
d.	40	35	104	55
e.	45	80	104	94

2. Which of the sets of data in Table 37–1 would you expect to find in a normal person?

   a   b   c   d   e

3. Which set of data would you expect in a person at rest breathing air at an altitude of 3000 m (10,000 feet; barometric pressure = 520 mm Hg)?

   a   b   c   d   e

4. Which set of data would you expect in a patient with a large left-to-right shunt?

a   b   c   d   e

5. Which set of data would you expect in a patient with a collapsed left lung?

a   b   c   d   e

6. Which set of data would you expect in a normal subject at rest breathing 100% O_2 in an unpressurized airplane cabin at 10,000 m (32,800 feet; barometric pressure = 180 mm Hg)?

a   b   c   d   e

The next 2 questions refer to Table 37–2.

Table 37–2. Arterial blood values.

	P_{CO_2} (mm Hg)	pH	HCO_3^- (meq/L)
a.	60	7.25	29
b.	25	7.30	12
c.	42	7.55	35
d.	25	7.50	20
e.	60	7.05	15

7. Which set of data would you expect in a man who is breathing through a tube that greatly increases his dead space?

a   b   c   d   e

8. Which set of data would you expect in a woman who has flown to an altitude of 4500 m (14,750 feet) in the open cockpit of an airplane?

a   b   c   d   e

In the following questions, one or more than one of the answers may be correct. Select—

A if (1), (2), and (3) are correct;
B if (1) and (3) are correct;
C if (2) and (4) are correct;
D if only (4) is correct; and
E if all are correct.

9. O_2 delivery from the blood to exercising muscle is facilitated by—
    (1) increased 2,3-DPG concentrations in red blood cells.
    (2) increased tissue temperature.
    (3) decreased tissue pH.
    (4) decreased mean distance from blood to cells.

A   B   C   D   E

10. Pulmonary fibrosis would be expected to produce—
(1) histotoxic hypoxia.
(2) hypoxic hypoxia.
(3) decreased vital capacity.
(4) cyanosis.

A   B   C   D   E

11. Deleterious effects of chronic cigarette smoking might be expected to include—
(1) patches of atelectasis.
(2) stagnant hypoxia.
(3) loss of elastic tissue in the lung.
(4) decreased anatomic dead space.

A   B   C   D   E

12. O_2 delivery to the tissues would be reduced in—
(1) a normal subject breathing air on top of Mt. Everest.
(2) a normal subject running a marathon at sea level.
(3) a patient with carbon monoxide poisoning.
(4) a patient who has ingested cyanide.

A   B   C   D   E

13. Which of the following are manifestations of oxygen toxicity?
(1) irritation of the respiratory tract.
(2) difficulty in seeing because of retrolental fibroplasia.
(3) convulsions.
(4) lung cysts in infants.

A   B   C   D   E

14. Which of the following would be expected to occur in the passengers of an airplane flying at 12,500 m (41,000 feet) when the pressurization of the cabin suddenly fails?
(1) a marked decrease in arterial P_{O_2}.
(2) air embolism.
(3) bubbles of gas in the blood.
(4) a marked increase in arterial P_{CO_2}.

A   B   C   D   E

In the following questions, indicate whether the item on the left is greater than **(G)**, the same as **(S)**, or less than **(L)** the item on the right.

15. CO_2 content of blood from exercising muscle. G S L CO_2 content of blood from resting muscle.

16. O_2 content of blood from exercising muscle. G S L O_2 content of blood from resting muscle.

17. pH of blood from exercising muscle. G S L pH of blood from resting muscle.

18. Blood lactate level during moderate exercise.   **G S L**   Blood lactate level during severe exercise.

19. Respiratory rate in a patient with hypoxia due to blood loss.   **G S L**   Respiratory rate in a patient with hypoxia due to carbon monoxide poisoning.

20. Respiratory rate in a patient with hypoxia due to cyanide poisoning.   **G S L**   Respiratory rate in a patient with hypoxia due to carbon monoxide poisoning.

21. Tidal volume produced by the supine pressure method of artificial respiration.   **G S L**   Tidal volume produced by mouth-to-mouth resuscitation.

22. Increase in O_2 content of arterial blood produced by breathing 100% oxygen in a subject with hypoxia due to a large atrial septal defect.   **G S L**   Increase in O_2 content of arterial blood produced by breathing 100% oxygen in a subject with hypoxia due to obstruction of a main branch of the pulmonary artery.

23. Ventilation after 4 hours' exposure to high altitude.   **G S L**   Ventilation after 4 days' exposure to high altitude.

24. Hematocrit of a subject who lives at an altitude of 550 m (1800 feet).   **G S L**   Hematocrit of a subject who lives at an altitude of 5500 m (18,000 feet).

25. Incidence of the bends when ascent from a dive is rapid.   **G S L**   Incidence of the bends when ascent from a dive is slow.

REFERENCES

Frohlich ED (editor): *Pathophysiology*, 3rd ed. Section II: Respiratory mechanisms, pp 119–202. Lippincott, 1984.

Renal Function & Micturition

38

Chapter 38 is concerned with the urinary system. The functional anatomy of the kidney is briefly reviewed along with the renal circulation. Glomerular filtration and the absorptive and secretory functions of the tubules are analyzed. Renal handling of water, Na^+, K^+, H^+, and other substances is reviewed; and brief consideration is given to the mechanisms of action of diuretics and the pathophysiology of renal disease. Finally, the normal function of the bladder is discussed and there is a brief review of the pathophysiology of bladder disease. The hormones produced by the kidneys are discussed in Chapter 24.

OBJECTIVES

The material in the chapter should help students to—

- Describe the morphology of a typical nephron and its blood supply.

- Define and give normal values for renal plasma flow and renal blood flow.

- Define autoregulation, and list the major theories advanced to explain autoregulation in the kidneys.

- Compare blood flow in the renal cortex to blood flow in the renal medulla, and list the major factors affecting renal blood flow.

- Outline the functions of the renal nerves.

- Define glomerular filtration rate, describe how it can be measured, and list the major factors affecting it.

- Discuss tubular reabsorption, using glucose as an example.

- Discuss tubular secretion, using K^+ as an example.

- Discuss tubular handling of Na^+.

- Summarize tubular handling of Cl^-, HCO_3^-, urea, and uric acid in terms of amounts filtered, secreted, reabsorbed, and excreted in urine.

- Describe how the countercurrent mechanism in the kidney operates to produce a hypertonic or hypotonic urine.

- Outline the processes involved in the secretion of H^+ into the tubules, and discuss the significance of these processes in the regulation of acid-base balance.

- List the major classes of diuretics and how each operates to increase urine flow.

- Describe the voiding reflex.

- List the major types of abnormal bladder function.

GENERAL QUESTIONS

1. Explain why, when the Tm of a substance that is secreted by the tubules is reached, clearance of the substance decreases as its plasma concentration increases.

2. What is the physiologic role of the mesangial cells in the glomeruli, and how do they carry it out?

3. Why does the clearance of glucose rise from zero and gradually approach the clearance of inulin as the plasma glucose level increases?

4. What happens to renal function when the kidneys swell within their relatively inelastic capsule?

5. What are the characteristics of a substance that can be used to measure the glomerular filtration rate (GFR)?

6. Discuss tubuloglomerular feedback from the point of view of the phenomenon itself, its physiologic function, and the possible mechanisms by which it is brought about.

7. Why is acidosis a common complication of chronic renal disease? How would you treat it?

8. On a suitable graph, draw the curves relating plasma levels to urinary excretion for inulin, glucose, para-aminohippurate (PAH), and bicarbonate. Explain each curve.

9. Compare the cellular mechanisms responsible for H^+ secretion in the proximal tubule with those in the distal tubule and those in the gastric mucosa.

10. Discuss the mechanisms responsible for the adaptation of NH_4^+ excretion that develops over a period of days in prolonged acidosis.

11. What is a cystometrogram? Describe and explain its components.

12. The following observations were made on a patient:

Plasma [HCO_3^-]	20 meq/L
GFR	125 mL/min
24-hour urine volume	1500 mL
Urinary [HCO_3^-]	25 meq/L
Urinary [NH_4^+]	75 meq/L
Urinary titratable acidity	50 meq/L

    A. Approximately how much HCO_3^- is being reabsorbed per 24 hours?
    B. How much Na^+ is being reabsorbed with the HCO_3^-?
    C. How much H^+ is being secreted by the renal tubules per 24 hours?

MULTIPLE-CHOICE QUESTIONS

In the following questions, select the single best answer.

1. In the presence of vasopressin, the greatest fraction of filtered water is absorbed in the—
    a. proximal tubule.
    b. loop of Henle.
    c. distal tubule.
    d. cortical collecting duct.
    e. medullary collecting duct.

2. In the absence of vasopressin, the greatest fraction of filtered water is absorbed in the—
   a. proximal tubule.
   b. loop of Henle.
   c. distal tubule.
   d. cortical collecting duct.
   e. medullary collecting duct.

In questions 3–7, match the disease or condition with the appropriate pattern of laboratory findings in Table 38–1.

Table 38–1.

	24-Hour Urine Volume	Ketones	Glucose	Protein
a.	1.4 L	+	0	0
b.	6.2 L	2+	4+	0
c.	1.6 L	0	0	4+
d.	6.4 L	0	0	0
e.	0.4 L	0	0	0

3. Diabetes insipidus   a b c d e

4. Nephrosis   a b c d e

5. Fasting   a b c d e

6. Dehydration   a b c d e

7. Diabetes mellitus   a b c d e

8. If the clearance of a substance which is freely filtered is less than that of inulin—
   a. there is net reabsorption of the substance in the tubules.
   b. there is net secretion of the substance in the tubules.
   c. the substance is neither secreted nor reabsorbed in the tubules.
   d. the substance becomes bound to protein in the tubules.
   e. the substance is secreted in the proximal tubule to a greater degree than it is in the distal tubule.

9. Glucose reabsorption occurs in the—
   a. proximal tubule.
   b. loop of Henle.
   c. distal tubule.
   d. cortical collecting duct.
   e. medullary collecting duct.

10. Aldosterone exerts its greatest effect on the—
    a. glomerulus.
    b. proximal tubule.
    c. thin portion of the loop of Henle.
    d. thick portion of the loop of Henle.
    e. collecting duct.

11. What is the clearance of a substance when its concentration in the plasma is 10 mg/dL, its concentration in the urine is 100 mg/dL, and urine flow is 2 mL/min?
  a. 2 mL/min.
  b. 10 mL/min.
  c. 20 mL/min.
  d. 200 mL/min.
  e. clearance cannot be determined from the information given.

12. The renal transport maximum (Tm) for a substance is—
  a. the maximum rate at which it can be filtered.
  b. the maximum rate at which it can be reabsorbed or secreted by the tubules.
  c. the maximum rate at which it can be excreted.
  d. the maximum degree to which it can be concentrated in the urine.
  e. the maximum degree to which it can be diluted in the urine.

13. As urine flow increases during osmotic diuresis—
  a. the osmolality of urine falls below that of plasma.
  b. the osmolality of urine increases because of the increased amounts of nonreabsorbable solute in the urine.
  c. the osmolality of urine approaches that of plasma because plasma leaks into the tubules.
  d. the osmolality of urine approaches that of plasma because an increasingly large fraction of the excreted urine is isotonic proximal tubular fluid.
  e. the action of vasopressin on the renal tubules is inhibited.

Questions 14, 15, and 16 relate to the data in Table 38–2, which were obtained in a normal woman:

Table 38–2.

	Control Period	Experimental Period
Arterial plasma		
Inulin (mg/mL)	0.004	0.004
Glucose (mg/dL)	100 (1 mg/mL)	300 (3 mg/mL)
Urea (μmol/mL)	5	5
Urine		
Inulin (mg/mL)	0.4	0.2
Glucose (mg/mL)	0	5
Urea (μmol/mL)	300	160
Urine flow (mL/min)	1	2

14. The glomerular filtration rate in the experimental period is—
  a. 4 times that in the control period.
  b. twice that in the control period.
  c. the same as in the control period.
  d. one-half that in the control period.
  e. one-quarter that in the control period.

15. The clearance of urea in the experimental period is—
  a. increased, probably because the urine flow is increased.
  b. increased, probably because the urine glucose concentration is increased.
  c. the same as in the control period.
  d. decreased, probably because the urine flow is increased.
  e. decreased, probably because the amount of urea filtered is decreased.

16. The Tm for glucose in this woman—
   a. is 100 mg/min.
   b. is 130 mg/min.
   c. is 200 mg/min.
   d. is 290 mg/min.
   e. cannot be calculated from the data given in Table 38–2.

In the following questions, one or more than one of the answers may be correct. Select—

> **A** if (1), (2), and (3) are correct;
> **B** if (1) and (3) are correct;
> **C** if (2) and (4) are correct;
> **D** if only (4) is correct; and
> **E** if all are correct.

17. Substances are removed from the tubular fluid by—
   (1) endocytosis.
   (2) active transport.
   (3) secondary active transport.
   (4) passive diffusion.

A   B   C   D   E

18. The kidneys secrete—
   (1) 1,25-dihydroxycholecalciferol.
   (2) renin.
   (3) erythropoietin.
   (4) vasopressin.

A   B   C   D   E

19. Which of the following hormones exert effects on the kidneys?
   (1) parathyroid hormone.
   (2) growth hormone.
   (3) cortisol.
   (4) luteinizing hormone.

A   B   C   D   E

20. Which of the following are quantitatively important in adding H^+ to the tubular fluid?
   (1) an Na^+-H^+ exchange pump.
   (2) diffusion of H^+ from the tubular cells into the tubular fluid.
   (3) an ATP-powered proton pump.
   (4) cotransport of Na^+ with H^+ into the tubular fluid.

A   B   C   D   E

21. An increase in the concentration of aldosterone in the blood would be expected to—
   (1) increase the amount of Na^+ reabsorbed from the tubular fluid.
   (2) increase the amount of H^+ secreted into the tubular fluid.
   (3) increase the amount of K^+ secreted into the tubular fluid.
   (4) decrease the urine volume.

A   B   C   D   E

22. Which of the following indicate the presence of disease in a man?
    (1) a PAH clearance of 650 mL/min.
    (2) an inulin clearance of 130 mL/min.
    (3) a urinary pH of 5.5.
    (4) a glucose clearance of 15 mL/min.

    A   B   C   D   E

23. Because a countercurrent multiplier system is present in the kidneys—
    (1) the glomerular filtration rate is held relatively constant.
    (2) blood flow in the vasa recta is slow.
    (3) the urine becomes acidified.
    (4) there is persistent hyperosmolality in the medullary pyramids.

    A   B   C   D   E

24. In a normal individual who drinks 1 L of 0.9% sodium chloride solution—
    (1) aldosterone secretion increases.
    (2) the urine volume increases.
    (3) the osmolality of the urine increases.
    (4) the amount of sodium in the urine increases.

    A   B   C   D   E

25. The bladder and its sphincters are innervated by—
    (1) sympathetic postganglionic neurons.
    (2) sensory fibers that traverse sympathetic pathways and the pudendal nerves.
    (3) parasympathetic preganglionic fibers in the sacral nerves.
    (4) somatic motor fibers in the hypogastric nerves.

    A   B   C   D   E

In questions 26–30, select the letter or letters identifying the appropriate part of the nephron in Fig 38–1. A lettered part may be selected once, more than once, or not at all.

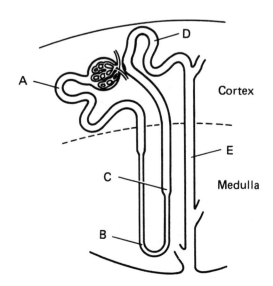

Figure 38–1. Juxtamedullary nephron.

26. The site or sites at which furosemide acts **A B C D E**

27. The site or sites at which thiazides act **A B C D E**

28. The site or sites at which tubular fluid osmolality exceeds that of plasma in the absence of vasopressin **A B C D E**

29. The site or sites at which Na⁺ is actively reabsorbed **A B C D E**

30. The site or sites at which K⁺ is secreted **A B C D E**

In the following questions, indicate whether the item on the left is greater than **(G)**, the same as **(S)**, or less than **(L)** the item on the right.

31. Percentage of cardiac output flowing through the kidneys. **G S L** Percentage of cardiac output flowing through the brain.

32. Clearance of inulin. **G S L** Clearance of albumin.

33. Glomerular filtration rate when the afferent arteriole is constricted. **G S L** Glomerular filtration rate when the efferent arteriole is constricted.

34. Maximal urine concentration on a low-protein diet. **G S L** Maximal urine concentration on a high-protein diet.

35. Hydrostatic pressure in the afferent arterioles. **G S L** Hydrostatic pressure in the efferent arterioles.

36. Amount of potassium in the urine after treatment with amiloride. **G S L** Amount of potassium in the urine after treatment with ethacrynic acid.

37. Osmolality of interstitial fluid around the tip of the loop of Henle during osmotic diuresis. **G S L** Osmolality of interstitial fluid around the tip of the loop of Henle during water diuresis.

38. Renal blood flow at rest. **G S L** Renal blood flow during exercise.

39. Na⁺,K⁺ ATPase content of cells in the thin portion of the ascending limb of the loop of Henle. **G S L** Na⁺,K⁺ ATPase content of cells in the thick portion of the ascending limb of the loop of Henle.

40. In metabolic acidosis, the pH of the urine at the end of the proximal tubule. **G S L** In metabolic acidosis, the pH of the urine at the end of the distal tubule.

REFERENCES

Sullivan LP, Grantham JJ: *Physiology of the Kidney*, 2nd ed. Lea & Febiger, 1982.
Vander AJ: *Renal Physiology*, 3rd ed. McGraw-Hill, 1985.

39

Regulation of Extracellular Fluid Composition & Volume

Chapter 39 is a review of the homeostatic mechanisms that operate to maintain the osmolality, volume, and ionic composition of the extracellular fluid within normal limits. The mechanisms that operate to maintain normal concentrations of H^+ in the body fluids are analyzed, and metabolic acidosis, metabolic alkalosis, respiratory acidosis, and respiratory alkalosis are considered in detail.

OBJECTIVES

The material in the chapter should help students to—

- Describe how the tonicity (osmolality) of the extracellular fluid is maintained by alterations in water intake and vasopressin secretion.

- Describe how the volume of the extracellular fluid is maintained by alterations in renin and aldosterone secretion.

- Name the mechanisms that operate to maintain the constancy of plasma concentrations of glucose, Ca^{2+}, and Na^+.

- Define acidosis and alkalosis, and give (in meq/L and pH) the normal mean and the range of H^+ concentrations in blood that are compatible with health.

- List the principal buffers in blood, interstitial fluid, and intracellular fluid; and, using the Henderson-Hasselbalch equation, describe what is unique about the bicarbonate buffer system.

- Describe the changes in blood chemistry that occur during the development of metabolic acidosis, and the respiratory and renal compensations for this condition.

- Describe the changes in blood chemistry that occur during the development of metabolic alkalosis, and the respiratory and renal compensations for this condition.

- Describe the changes in blood chemistry that occur during the development of respiratory acidosis, and the renal compensation for this condition.

- Describe the changes in blood chemistry that occur during the development of respiratory alkalosis, and the renal compensation for this condition.

- Summarize the effects of changes in potassium metabolism on acid-base balance.

GENERAL QUESTIONS

1. Compare and contrast the mechanisms that operate to maintain the osmolality of the extracellular fluid and those that operate to maintain its volume. How do they differ, and how do they overlap?

2. What are the main sources of the acid loads presented to the body in everyday living? What are some common diseases that cause increased acid loads in the body, and in each, how is the load produced?

3. Compare and contrast metabolic acidosis and respiratory acidosis.

4. What conditions might be present in patients with the following blood chemistry values, and why?
   a. pH, 7.28; HCO_3^-, 18.1 meq/L; P_{CO_2}, 40 mm Hg.
   b. pH, 7.34; HCO_3^-, 25 meq/L; P_{CO_2}, 48 mm Hg.

5. Why does excess secretion of vasopressin, the water-regulating hormone, cause a decrease in plasma Na^+, whereas excess secretion of aldosterone, the salt-regulating hormone, causes no change or only a small increase in plasma Na^+?

6. Describe and explain the immediate and long-term changes in body chemistry and function you would expect to see in an otherwise normal individual who has indigestion and ingests a large amount of $NaHCO_3$. What immediate changes would occur in the extracellular fluid volume and body weight of this person?

7. Describe and explain the alterations in extracellular fluid volume and acid-base balance that occur in patients with chronically elevated plasma aldosterone concentrations due to primary hyperaldosteronism.

8. Describe by means of a diagram plotting plasma HCO_3^- concentration against pH (a "Davenport diagram") the immediate and more long-term changes that occur in the acid-base balance of a normal individual who hyperventilates for 5 minutes.

9. Why do the K^+ and H^+ concentrations in extracellular fluid generally parallel each other?

10. In prolonged respiratory acidosis, changes in urinary pH, titratable acidity, HCO_3^- concentration, NH_4^+ concentration, and K^+ concentration would be expected. In which direction would each change, and why?

MULTIPLE-CHOICE QUESTIONS

In the following questions, select the single best answer.

Questions 1–4 relate to the numbered points on the nomogram in Fig 39–1.

1. Point 1 identifies values that would be seen—
   a. in a mountain climber after several weeks at high altitude.
   b. in long-standing pulmonary disease.
   c. in diabetic coma.
   d. after 5 minutes of hyperventilation.
   e. after prolonged vomiting.

2. Point 2 identifies values that would be seen—
   a. in a mountain climber after several weeks at high altitude.
   b. in long-standing pulmonary disease.
   c. in diabetic coma.
   d. after 5 minutes of hyperventilation.
   e. after prolonged vomiting.

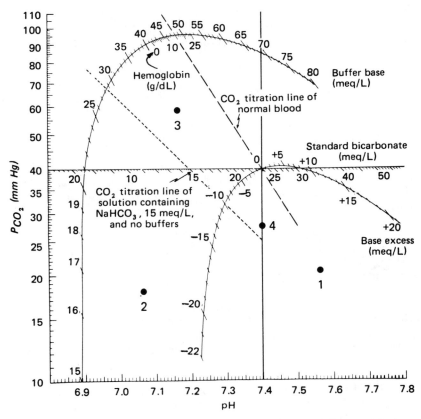

Figure 39–1. Siggaard-Andersen curve nomogram. (Courtesy of O Siggaard-Andersen and Radiometer, Copenhagen, Denmark.)

3. Point 3 identifies values that would be seen—
    a. in a mountain climber after several weeks at high altitude.
    b. in long-standing pulmonary disease.
    c. in diabetic coma.
    d. after 5 minutes of hyperventilation.
    e. after prolonged vomiting.

4. Point 4 identifies values that would be seen—
    a. in a mountain climber after several weeks at high altitude.
    b. in long-standing pulmonary disease.
    c. in diabetic coma.
    d. after 5 minutes of hyperventilation.
    e. after prolonged vomiting.

5. Increasing alveolar ventilation increases the blood pH because—
    a. it activates neural mechanisms that remove acid from the blood.
    b. it makes hemoglobin a stronger acid.
    c. it increases the P_{O_2} of the blood.
    d. it decreases the P_{CO_2} in the alveoli.
    e. the increased muscle work of increased breathing generates more CO_2.

6. In uncompensated respiratory alkalosis—
    a. the plasma pH, the plasma HCO_3^- concentration, and the arterial P_{CO_2} are all low.
    b. the plasma pH is high and the plasma HCO_3^- concentration and the arterial P_{CO_2} are low.

    c. the plasma pH and the plasma HCO_3^- concentration are low and the arterial Pco_2 is normal.

    d. the plasma pH is high, the plasma HCO_3^- concentration is low, and the arterial Pco_2 is normal.

    e. the plasma pH is low, the plasma HCO_3^- concentration is high, and the arterial Pco_2 is normal.

7. When the pH of a plasma sample is 7.62, its H^+ concentration, in nanoequivalents per liter (neq/L), is—

    a. 41.

    b. 410.

    c. 240.

    d. 24.

    e. 2.4.

8. In a patient with a plasma pH of 7.10, the $[HCO_3^-]/[H_2CO_3]$ ratio in plasma is—

    a. 20.

    b. 10.

    c. 2.

    d. 1.

    e. 0.1.

9. In a patient who has become dehydrated, body water should be replaced by intravenous infusion of—

    a. distilled water.

    b. 0.9% sodium chloride solution.

    c. 5% glucose solution.

    d. hyperoncotic albumin.

    e. 10% glucose solution.

In the following questions, one or more than one of the answers may be correct. Select—

          A if (1), (2), and (3) are correct;
          B if (1) and (3) are correct;
          C if (2) and (4) are correct;
          D if only (4) is correct; and
          E if all are correct.

10. The plasma concentration of HCO_3^- is low in—

    (1) compensated metabolic acidosis.

    (2) uncompensated metabolic acidosis.

    (3) compensated respiratory alkalosis.

    (4) uncompensated respiratory acidosis.

          A   **B**   **C**   **D**   **E**

11. Sudden expansion of the extracellular fluid volume due to infusion of isotonic saline solution would be expected to cause—

    (1) decreased secretion of vasopressin.

    (2) increased secretion of atrial natriuretic peptide.

    (3) decreased secretion of renin.

    (4) decreased secretion of aldosterone.

          A   **B**   **C**   **D**   **E**

12. The plasma osmolality decreases after infusion of—
     (1) isotonic saline solution.
     (2) vasopressin.
     (3) aldosterone.
     (4) isotonic glucose solution.

                    A   B   C   D   E

13. If acetazolamide, a carbonic anhydrase inhibitor, could be administered in
     a dose that blocked carbonic anhydrase in the renal tubular cells without
     affecting it in other cells in the body, one would expect to see—
     (1) a more acidic urine.
     (2) a decrease in blood pH.
     (3) a decrease in Na^+ secretion.
     (4) an increase in urinary HCO_3^- concentration.

                    A   B   C   D   E

14. If the amount of a strong acid produced by metabolism increases from 60
     meq/d to 90 meq/d—
     (1) the amount of free H^+ excreted in the urine will increase.
     (2) the amount of NH_4^+ excreted in the urine will increase.
     (3) the titratable acidity of the urine will increase.
     (4) the blood pH will not change or will decrease slightly.

                    A   B   C   D   E

15. Metabolism of which of the following kinds of amino acids results in a
     significant increase in the acid load presented for buffering and excretion?
     (1) aromatic amino acids.
     (2) sulfur-containing amino acids.
     (3) dibasic amino acids.
     (4) phosphorylated amino acids.

                    A   B   C   D   E

In the following questions, indicate whether the item on the left is greater than
(G), the same as (S), or less than (L) the item on the right.

16. pH of blood in metabolic   G S L   pH of blood in metabolic
     acidosis.   alkalosis.

17. pH of arterial blood supply-   G S L   pH of venous blood coming
     ing skeletal muscle.   from skeletal muscle.

18. Na^+ concentration in extra-   G S L Na^+ concentration in intra-
     cellular fluid.   cellular fluid.

19. Cl^- concentration in extra-   G S L Cl^- concentration in intra-
     cellular fluid.   cellular fluid.

20. K^+ concentration in extra-   G S L K^+ concentration in intra-
     cellular fluid.   cellular fluid.

21. Na$^+$ concentration in nor-   **G S L**   Na$^+$ concentration in iso-
    mal plasma.   tonic saline.

22. Osmolality of extracellular   **G S L**   Osmolality of extracellular
    fluid before drinking 1 L of   fluid after drinking 1 L of
    isotonic saline solution.   isotonic saline solution.

23. Urinary NH$_4^+$ concentra-   **G S L**   Urinary NH$_4^+$ concentra-
    tion when the arterial P$_{CO_2}$ tion when the arterial P$_{CO_2}$
    is chronically elevated.   is chronically depressed.

24. Arterial P$_{CO_2}$ in uncompen-   **G S L**   Arterial P$_{CO_2}$ in uncompen-
    sated metabolic acidosis.   sated metabolic alkalosis.

25. Acid secretion by renal   **G S L**   Acid secretion by renal
    tubular cells when body K$^+$ tubular cells when body K$^+$
    is depleted.   is greater than normal.

REFERENCES

Davenport HW: *The ABC of Acid-Base Chemistry*, 6th ed. University of Chicago Press, 1974.
Rose BD: *Clinical Physiology of Acid-Base and Electrolyte Disorders*. McGraw-Hill, 1984.

Answers to Multiple-Choice Questions

CHAPTER 1: THE GENERAL & CELLULAR BASIS OF MEDICAL PHYSIOLOGY

Single best answer

1. c	2. e	3. d	4. a	5. c	6. e

One or more than one

7. C	8. C	9. A	10. B	11. E	12. B
13. E	14. E				

Matching questions

15. B	16. A	17. C, D	18. C

G, S, L

19. G	20. G	21. L	22. S	23. L	24. G
25. G	26. S	27. L			

CHAPTER 2: EXCITABLE TISSUE: NERVE

Single best answer

1. b	2. a	3. b	4. e	5. b	6. e

One or more than one

7. E	8. B	9. E	10. A	11. B

G, S, L

12. G	13. L	14. G	15. L	16. L	17. L
18. G	19. L	20. L			

CHAPTER 3: EXCITABLE TISSUE: MUSCLE

Single best answer

1. b	2. d	3. e	4. c	5. c	6. b
7. e	8. d				

Matching questions

9. A	10. C	11. B	12. D	13. C

One or more than one

14. A	15. E	16. C	17. E	18. E	19. E

G, S, L

20. G	21. G	22. L	23. G	24. L	25. L

CHAPTER 4: SYNAPTIC & JUNCTIONAL TRANSMISSION

Single best answer

1. d	2. c	3. e	4. e	5. d	6. c
7. d					

One or more than one
8. A   9. E   10. D   11. E   12. D

G, S, L
13. L   14. G   15. L   16. S   17. G   18. S
19. G   20. L   21. L   22. L

CHAPTER 5: INITIATION OF IMPULSES IN SENSE ORGANS

Single best answer
1. c   2. d   3. d

One or more than one
4. A   5. E   6. B   7. A   8. E

G, S, L
9. L   10. S   11. L   12. S   13. G

CHAPTER 6: REFLEXES

Single best answer
1. c   2. e   3. d

One or more than one
4. C   5. E   6. B   7. C   8. A   9. B
10. A

G, S, L
11. S   12. G   13. G   14. G   15. L   16. L
17. G   18. G   19. L   20. L

CHAPTER 7: CUTANEOUS, DEEP, & VISCERAL SENSATION

Single best answer
1. d   2. a   3. c   4. c   5. d

Matching questions
6. C   7. A   8. C   9. C   10. B

One or more than one
11. A   12. C   13. A   14. D   15. B

G, S, L
16. L   17. L   18. L   19. S   20. L   21. G
22. L   23. G

CHAPTER 8: VISION

Single best answer
1. d   2. d   3. e   4. b   5. e   6. d

One or more than one
7. E   8. C   9. E   10. E   11. D   12. C

G, S, L
13. G   14. L   15. G   16. L   17. L   18. L
19. G   20. L   21. G

CHAPTER 9: HEARING & EQUILIBRIUM

Single best answer
1. d   2. e   3. d   4. c   5. d

Matching questions
6. A   7. D   8. B   9. B   10. A

One or more than one
11. A   12. E   13. A   14. D   15. A

G, S, L
16. L   17. S   18. L   19. L   20. G   21. L
22. G   23. L   24. L   25. G

CHAPTER 10: SMELL & TASTE

Single best answer
1. d   2. a   3. a   4. e

One or more than one
5. B   6. E   7. C   8. A   9. C

G, S, L
10. G   11. L   12. L   13. G   14. G   15. G

CHAPTER 11: AROUSAL MECHANISMS, SLEEP, & THE ELECTRICAL ACTIVITY OF THE BRAIN

Single best answer
1. c   2. e   3. c   4. d

Matching questions
5. B   6. A   7. A   8. B   9. B

One or more than one
10. D   11. A   12. A   13. B   14. C

G, S, L
15. L   16. L   17. L   18. L   19. G   20. G
21. L

CHAPTER 12: CONTROL OF POSTURE & MOVEMENT

Single best answer
1. c   2. b   3. e   4. d

One or more than one
5. E   6. B   7. A   8. C   9. B

Matching questions
10. A,B,C   11. E   12. none   13. A   14. none   15. B
16. C   17. A   18. C,D

G, S, L
19. G   20. G   21. L   22. L   23. L   24. G

CHAPTER 13: THE AUTONOMIC NERVOUS SYSTEM

Single best answer
1. c   2. e

One or more than one
3. A   4. C   5. D   6. C   7. E   8. E

G, S, L
9. L   10. G   11. G   12. G   13. L   14. L
15. G   16. G

CHAPTER 14: CENTRAL REGULATION OF VISCERAL FUNCTION

Single best answer
1. b   2. d   3. e

Matching questions
4. B,C   5. C   6. C   7. E   8. D   9. B
10. A   11. A   12. B   13. D

One or more than one
14. B   15. E   16. D   17. A   18. B   19. A
20. C

G, S, L
21. G   22. L   23. G   24. L   25. L

CHAPTER 15: NEURAL BASIS OF INSTINCTUAL BEHAVIOR & EMOTIONS

Single best answer
1. B   2. A   3. A   4. D   5. A   6. B

One or more than one
7. C   8. A   9. E   10. E   11. B   12. D
13. E

G, S, L
14. L   15. L   16. G   17. L   18. L   19. G

CHAPTER 16: CONDITIONED REFLEXES, LEARNING & RELATED PHENOMENA

Single best answer
1. c   2. a   3. e   4. c   5. d

One or more than one
6. A   7. C   8. A   9. C   10. D

Matching questions
11. B   12. A,C   13. B   14. B,D   15. C   16. B

G, S, L
17. L   18. L   19. L   20. G   21. G   22. L
23. L   24. G

CHAPTER 17: ENERGY BALANCE, METABOLISM, & NUTRITION

Single best answer

1. b   2. a   3. d   4. d   5. e   6. b
7. c

One or more than one

8. E   9. E   10. A   11. C   12. C   13. B
14. D   15. B

G, S, L

16. G   17. L   18. G   19. G   20. G   21. G
22. G   23. L   24. G   25. L   26. L   27. G
28. L   29. L   30. G

CHAPTER 18: THE THYROID GLAND

Single best answer

1. d   2. d   3. e   4. a   5. c   6. d
7. b

One or more than one

8. E   9. B   10. B   11. C   12. C   13. D
14. C   15. A

G, S, L

16. G   17. L   18. L   19. G   20. L   21. G
22. S   23. G   24. G   25. G

CHAPTER 19: ENDOCRINE FUNCTIONS OF THE PANCREAS & THE REGULATION OF CARBOHYDRATE METABOLISM

Single best answer

1. e   2. d   3. e   4. d   5. a   6. d
7. e   8. c   9. d   10. c

One or more than one

11. A   12. E   13. C   14. A   15. E   16. E
17. E   18. A   19. A   20. E

G, S, L

21. L   22. G   23. L   24. L   25. L   26. L
27. G   28. G   29. L   30. G

CHAPTER 20: THE ADRENAL MEDULLA & ADRENAL CORTEX

Single best answer

1. d   2. c   3. e   4. d   5. b   6. e
7. c   8. b   9. d   10. c

One or more than one

11. A   12. B   13. A   14. C   15. D   16. B
17. C   18. E

Matching questions

19. E   20. A, B   21. A   22. D   23. C,D

G, S, L
24. G   25. G   26. G   27. L   28. G   29. G
30. G   31. L   32. G   33. G

CHAPTER 21: HORMONAL CONTROL OF CALCIUM METABOLISM & THE PHYSIOLOGY OF BONE

Single best answer
1. c   2. e   3. d   4. e   5. a

One or more than one
6. E   7. A   8. E   9. C   10. E

G, S, L
11. L   12. G   13. G   14. G   15. G   16. L

CHAPTER 22: THE PITUITARY GLAND

Single best answer
1. e   2. d   3. d   4. e

One or more than one
5. A   6. A   7. E   8. E   9. A   10. B

G, S, L
11. L   12. G   13. S   14. L   15. G   16. L

CHAPTER 23: THE GONADS: DEVELOPMENT & FUNCTION OF THE REPRODUCTIVE SYSTEM

Single best answer
1. c   2. e   3. d   4. a   5. a   6. e
7. e   8. e   9. a   10. e   11. c   12. a
13. b   14. d

One or more than one
15. A   16. E   17. B   18. A   19. E   20. C
21. E   22. C   23. C   24. E   25. E   26. D
27. E   28. A

G, S, L
29. L   30. S   31. G   32. L   33. G   34. L
35. G   36. G   37. G   38. L

CHAPTER 24: OTHER ENDOCRINE ORGANS

Single best answer
1. d   2. c   3. d   4. d   5. e

One or more than one
6. C   7. E   8. D   9. D   10. E

G, S, L
11. G   12. G   13. L   14. G   15. L   16. G

CHAPTER 25: DIGESTION & ABSORPTION

Single best answer

1. b	2. c	3. c	4. e	5. b

One or more than one

6. E	7. C	8. E	9. E	10. B

G, S, L

11. G	12. G	13. S	14. G	15. G	16. G
17. L	18. G	19. L	20. S		

CHAPTER 26: REGULATION OF GASTROINTESTINAL FUNCTION

Single best answer

1. d	2. c	3. b	4. d	5. c	6. b
7. a	8. d	9. c	10. e	11. c	12. d
13. b	14. a				

One or more than one

15. A	16. E	17. C	18. E	19. C	20. A
21. D	22. D	23. C	24. B	25. E	26. B
27. E	28. D				

G, S, L

29. G	30. G	31. L	32. L	33. L	34. S
35. S	36. L				

CHAPTER 27: CIRCULATING BODY FLUIDS

Single best answer

1. b	2. a	3. a	4. e	5. c	6. c
7. e					

One or more than one

8. C	9. A	10. E	11. B	12. A	13. A
14. B					

Matching questions

15. B,C,D	16. E	17. E	18. A,E	19. A,C

G, S, L

20. L	21. S	22. L	23. L	24. L	25. G

CHAPTER 28: ORIGIN OF THE HEARTBEAT & THE ELECTRICAL ACTIVITY OF THE HEART

Single best answer

1. d	2. b	3. c	4. d	5. a	6. a
7. d	8. c	9. d			

One or more than one

10. B	11. D	12. A	13. B	14. A

G, S, L

15. L	16. L	17. G	18. L	19. L

CHAPTER 29: THE HEART AS A PUMP

Single best answer
1. d   2. c   3. e   4. c   5. c   6. e
7. c

One or more than one
8. C   9. C   10. A   11. B   12. C

G, S, L
13. G   14. G   15. G   16. G   17. L   18. G
19. L   20. L   21. G

CHAPTER 30: DYNAMICS OF BLOOD & LYMPH FLOW

Single best answer
1. d   2. a   3. e   4. c   5. c   6. c
7. b   8. b   9. d

One or more than one
10. B   11. E   12. D   13. E   14. C   15. E

G, S, L
16. L   17. G   18. G   19. L   20. G   21. G
22. L   23. G   24. L   25. L

CHAPTER 31: CARDIOVASCULAR REGULATORY MECHANISMS

Single best answer
1. e   2. e   3. c   4. b   5. d

One or more than one
6. C   7. A   8. D   9. E   10. D

G, S, L
11. G   12. L   13. G   14. G   15. L   16. G
17. L   18. G

CHAPTER 32: CIRCULATION THROUGH SPECIAL REGIONS

Single best answer
1. c   2. d   3. d   4. b   5. a   6. d
7. a   8. e   9. e   10. c

One or more than one
11. D   12. C   13. A   14. B   15. A

G, S, L
16. S   17. G   18. L   19. L   20. L   21. L
22. G   23. L   24. G   25. L   26. G

CHAPTER 33: CARDIOVASCULAR HOMEOSTASIS IN HEALTH & DISEASE

General Question 11
Cardiac output 4.73 L/min, stroke volume 73.9 mL.